高等学校测绘工程系列教材

（第三版）
测绘工程专业英语
English for Geomatics Engineering
（The Third Edition）

主编 尹晖 张晓鸣 王建国

WUHAN UNIVERSITY PRESS
武汉大学出版社

图书在版编目(CIP)数据

测绘工程专业英语/尹晖,张晓鸣,王建国主编．—3 版.—武汉:武汉大学出版社,2024.7
高等学校测绘工程系列教材
ISBN 978-7-307-24246-3

Ⅰ．测⋯　Ⅱ．①尹⋯　②张⋯　③王⋯　Ⅲ．工程测量—英语—高等学校—教材　Ⅳ．TB22

中国国家版本馆 CIP 数据核字(2024)第 028023 号

责任编辑:鲍　玲　　责任校对:汪欣怡　　版式设计:马　佳

出版发行:**武汉大学出版社**　(430072　武昌　珞珈山)
（电子邮箱:cbs22@whu.edu.cn　网址:www.wdp.com.cn）
印刷:武汉科源印刷设计有限公司
开本:787×1092　1/16　印张:15.5　字数:384 千字
版次:2004 年 12 月第 1 版　2013 年 8 月第 2 版
　　2024 年 7 月第 3 版　2024 年 7 月第 3 版第 1 次印刷
ISBN 978-7-307-24246-3　　定价:45.00 元

版权所有,不得翻印;凡购买我社的图书,如有质量问题,请与当地图书销售部门联系调换。

前 言

《测绘工程专业英语》教材自 2005 年第一版和 2013 年第二版发行以来，已重印二十多次，得到全国各兄弟院校和武汉大学测绘工程本科专业的广泛采用和认可，教学实践证明该教材编排合理、内容全面，是一本帮助测绘人员快速提高专业英语能力的实用教材。为此，我们在第三版编排中，继续保留了原教材基础部分的内容和风格。考虑到测绘新技术和高等教育国际化的迅速发展，第三版教材对部分内容作了较大的调整和修改，主要包括以下几点：

（1）Part I 由 20 个单元组成，保留了第二版测绘基础知识的全部内容，并新增了一个关于隧道测量及施工方法的单元。

（2）Part II 的内容重新进行了改编和整理，压缩为 6 个单元。主要涉及全球导航卫星系统、地理信息系统、数字地面模型、摄影测量、遥感和无人机测图技术的基础知识。

（3）Part III 为 4 个单元，由原教材 Part III 和 Part IV 的部分内容整合而成，主要涉及国际学术交流的基本知识点，如国际测绘学术组织、著名测绘期刊和专业英语论文写作及科技英语翻译技巧。

（4）Part IV 由 6 个单元组成，考虑到测绘工程专业学生本科毕业后出国深造的需要，新增了留学申请指南、履历/简历写作、求职信及与相关作文写作的方法、技巧及参考范文。

本教材修订工作由武汉大学尹晖教授组织，负责编写 Part I、Part III、Part IV 部分和 Unit 22、Unit 23 及全书的设计和统稿。Unit 24、Unit 25、Unit 26 由华东交通大学张晓鸣博士编写。Unit 19、Unit 21 由尹晖教授和加拿大约克大学王建国教授共同完成。

感谢全国各兄弟院校、武汉大学测绘学院同行提出的宝贵意见以及给予的大力支持，感谢武汉大学测绘学院部分研究生、本科生以及海外留学生的大力协助，限于我们的水平，书中不当之处恳请读者批评指正。

编　者
2024 年 1 月

Table of Contents

Part I Basic Knowledge in Geomatics Engineering 1
  Unit 1   What Is Geomatics? 3
  Unit 2   Geodetic Surveying and Plane Surveying 8
  Unit 3   Distance Measurement 13
  Unit 4   Angle and Direction Measurement 18
  Unit 5   Traversing 23
  Unit 6   Methods of Elevation Determination 28
  Unit 7   Robotic Total Station 33
  Unit 8   Errors in Measurement 37
  Unit 9   Basic Statistical Analysis of Random Errors 41
  Unit 10 Accuracy and Precision 46
  Unit 11 Least-Squares Adjustment 49
  Unit 12 Geodesy Concepts 53
  Unit 13 Geoid and Reference Ellipsoid 57
  Unit 14 Datums, Coordinates and Conversions 61
  Unit 15 Map Projection 65
  Unit 16 Gravity Measurement 70
  Unit 17 Optimal Design of Geomatics Network 74
  Unit 18 Construction Layout 79
  Unit 19 Tunnel Surveying and Construction Methods 83
  Unit 20 Deformation Monitoring of Engineering Structure 89

Part II Advanced Technologies in Geomatics Engineering 95
  Unit 21 GNSS Fundamentals 97
  Unit 22 GIS Basics 106
  Unit 23 Digital Terrain Modeling 111
  Unit 24 Photogrammetry Overview 116
  Unit 25 Fundamentals of Remote Sensing 122
  Unit 26 UAV Mapping Technology 129

Part III Basic International Academic Communication 135
  Unit 27 International Geoscience Organizations 137
  Unit 28 Prestigious Journals in Geomatics 142

	Unit 29	Professional English Paper Writing	149
	Unit 30	Translation Techniques for EST	158

Part IV Study Abroad Application Guide ... 165

	Unit 31	International Universities/Schools in Geomatics	167
	Unit 32	Get Ready to Study Abroad	173
	Unit 33	How to Write a CV	179
	Unit 34	Tips for Resume Writing	184
	Unit 35	Cover Letter Writing	189
	Unit 36	Essays Related to University Application	194

Appendix I Vocabulary ... 199

Appendix II Glossary of Terms ... 220

References ... 236

Part I

Basic Knowledge in Geomatics Engineering

Unit 1 What Is Geomatics?

Geomatics Defined

Where does the word Geomatics come from? GEODESY + GEOINFORMATICS = GEOMATICS or GEO- for Earth and -MATICS for mathematical or GEO- for Geoscience and -MATICS for informatics. It has been said that geomatics is many things to many people. The term geomatics emerged first in Canada and as an academic discipline. It has been introduced worldwide in a number of institutes of higher education during the past few decades, mostly by renaming what was previously called "geodesy" or "surveying", and by adding a number of computer science—and/ or GIS-oriented courses.[①] Now the term includes the traditional surveying definition along with surveying steadily increased importance with the development of new technologies and the growing demand for a variety of spatially related types of information, particularly in measuring and monitoring our environment. Increasingly critical are areas of expanding populations, appreciating land values, dwindling natural resources, and the continuing stressing of the quality of our land, water and air from human activities. As such, geomatics bridges wide arcs from the geosciences through various engineering sciences and computer sciences to spatial planning, land development and the environmental sciences. Now the word geomatics has been adopted by several international bodies including the International Standards Organization (ISO), so it is here to stay.

The term "surveyor" is traditionally used to collectively describe those engaged in the above activities. More explicit job descriptions such as Land Surveyor, Engineering Surveyor or Hydrographic Surveyor for example, are commonly used by practitioners to more clearly describe and market their specialized expertise.

The term geomatics is a recent creation to convey the true collective and scientific nature of these related activities and has the flexibility to allow for the incorporation of future technological developments in these fields. Adoption of the term also allows a coherent marketing of the profession to industry and schools on a worldwide basis.[②] As a result, both course and award titles in the traditional Land Surveying sector at many of the world's leading universities are being changed to "Degree in Geomatics". This does not suggest the demise of the term "surveyor" and graduates will still practice as land surveyors or photogrammetrists, etc., as appropriate to their specialization.

In the last decade, there has been dramatic development and growth in the use of hardware and software solutions to both measure and process geo-spatial data. This has created and will continue to create new areas of application, with associated job opportunities for suitably qualified graduates. As a result, the role of the "surveyor" is expanding beyond traditional areas of practice, as described above, into new areas of opportunity. In addition, recent advances in the technology of data collection and processing have blurred the boundaries of practice and activity between what were previously regarded as related but separate areas. Such developments are forecast to continue and will create new career paths for graduates whose education and training is broadly based and of a high academic standard.

To enable graduates to take full advantage of these developments, significant changes in education and training are required. Academic and professional institutions are also responding, in part, by adopting the term geomatics both as a course and as an award title. A working definition of geomatics, which reflects current thinking and predicted change, is:

The science and technology of acquiring, storing, processing, managing, analyzing and presenting geographically referenced information (geo-spatial data). This broad term applies both to science and technology, and integrates the following more specific disciplines and technologies including surveying and mapping, geodesy, satellite positioning, photogrammetry, remote sensing, geographic information systems (GIS), land management, computer systems, environmental visualization and computer graphics.

Several terms such as "geomatics", "geomatic engineering", and "geoinformatics" are now in common use pertaining to activities generally concerned with geographic information. These terms have been adopted primarily to represent the general approach that geographic information is collected, managed, and applied. Along with land surveying, photogrammetry, remote sensing, and cartography, GIS is an important component of geomatics.

Branches of Geomatics

Data acquisition techniques include field surveying, global positioning system (GPS), satellite positioning, and remotely sensed imagery obtained through aerial photography and satellite imagery. It also includes the acquisition of database material scanned from older maps and plans and data collected by related agencies.

Data management and process are handled through the use of computer programs for engineering design, digital photogrammetry, image analysis, relational data base management, and geographic information systems (GIS). Data plotting (presentation) is handled through the use of mapping and other illustrative computer programs; the presentations are displayed on computer screens (where interactive editing can occur) and are output on paper from digital plotting devices.

Once the positions and attributes of geographic entities have been digitized and stored in computer memory, they are available for use by a wide variety of users. Through the use of modern information technology (IT), geomatics brings together professionals in the following disciplines: surveying, mapping, remote sensing, land registration, civil and marine engineering, forestry, agriculture, planning and development, geology, geographical sciences, infrastructure management, navigation, environmental and natural resources monitoring, and computer science.

Other Definitions of Geomatics

<u>As defined by the Canadian Institute of Geomatics in their quarterly journal *Geomatica*: **Geomatics** is a field of activities which, using a systemic approach, integrates all the means used to acquire and manage spatial data required as part of scientific, administrative, legal and technical operations involved in the process of the production and management of spatial information.</u>③

The definition of **Geomatics** is evolving. A working definition might be "the art, science and technologies related to the management of geographically-referenced information". Geomatics includes a wide range of activities, from the acquisition and analysis of site-specific spatial data in engineering and development surveys to the application of GIS and remote sensing technologies in environmental management. It includes cadastral surveying, hydrographic surveying, and ocean mapping, and it plays an important role in land administration and land use management.

Geomatics is the modern scientific term referring to the integrated approach of measurement, analysis, management, storage and display of the descriptions and location of Earth-based data, often termed spatial data. These data come from many sources, including earth orbiting satellites, air and sea-borne sensors and ground based instruments. It is processed and manipulated with state-of-the-art information technology using computer software and hardware. It has applications in all disciplines which depend on spatial data, including environmental studies, planning, engineering, navigation, geology and geophysics, oceanography, land development and land ownership and tourism. It is thus fundamental to all the geoscience disciplines which use spatially related data.

[from the School of Geomatic Engineering, Univ. of New South Wales]

Geomatics is concerned with the measurement, representation, analysis, management, retrieval and display of spatial data concerning both the Earth's physical features and the built environment. The principal disciplines embraced by Geomatics include the mapping sciences, land management, geographic information systems, environmental visualisation, geodesy, photogrammetry, remote sensing and surveying.

[from the Dept. of Geomatics at Univ. of Melbourne]

Geomatics comprises the science, engineering, and art involved in collecting and managing geographically-referenced information. Geographical information plays an important role in activities such as environmental monitoring, management of land and marine resources, and real estate transactions.

[from the Dept. of Geodesy and Geomatics Engineering at UNB]

The science of **Geomatics** is concerned with the measurement, representation, analysis, management, retrieval and display of spatial information describing both the Earth's physical features and the built environment. Geomatics includes disciplines such as: Surveying, Geodesy, Remote Sensing & Photogrammetry, Cartography, Geographic Information Systems, Global Positioning Systems.

[from the Dept. of Surveying and Spatial Information Science at the Univ. of Tasmania]

Words and Expressions

geoscience [ˌdʒiːəʊˈsaɪəns] n. 地球科学
informatics [ˌɪnfəˈmætɪks] n. 信息学；情报学
monitor [ˈmɒnɪtə(r)] vt. 监控 n. 监测；监视；控制；追踪；监控器
appreciate [əˈpriːʃieɪt] vi. 增值；涨价 vt. 赏识；鉴赏；感激
dwindle [ˈdwɪndl] v. 缩小
ISO abbr. International Standardization Organization 国际标准化组织
explicit [ɪkˈsplɪsɪt] adj. 清楚的；外在的；坦率的；(租金等)直接付款的
hydrographic [haɪˈdrɒɡrəfɪk] adj. 与水道测量有关的；与水文地理有关的
hydrographic survey 海道测量；水道测量
practitioner [prækˈtɪʃənər] n. 从业者；开业者
expertise [ˈekspɜːtɪz] n. 专门技术；专家的意见
flexibility [flekˈsɪbɪlɪtɪ] n. 适应性；机动性；挠性
incorporation [ɪnˌkɔːpəˈreɪʃn] n. 结合；合并；形成法人组织；组成公司(或社团)
coherent [kəʊˈhɪərənt] adj. 一致的；连贯的
demise [dɪˈmaɪz] n. 死亡；让位；禅让 vt. 让渡；遗赠；转让
blur [blɜːr] v. 把(界线；视线等)弄得模糊不清；涂污；污损(名誉等)；弄污
visualization [ˌvɪʒʊəlaɪˈzeɪʃn] n. 可视化；清楚地呈现
pertaining [pə(ː)ˈteɪnɪŋ] adj. 与……有关系的；附属……的；为……固有的(to)
imagery [ˈɪmɪdʒəri] n. 肖像(总称)；雕刻影像
plotting [ˈplɒtɪŋ] n. 标图；测绘
illustrative [ɪˈlʌstrətɪv] adj. 说明性的；例证性的
entity [ˈentɪti] n. 实体
digitize [ˈdɪdʒɪtaɪz] v. [计]将资料数字化
registration [ˌredʒɪˈstreɪʃn] n. 注册；报到；登记
forestry [ˈfɒrɪstri] n. 林产；森林地；林学
geology [dʒiˈɒlədʒi] n. 地质学；地质概况
geographical [ˌdʒiːəˈɡræfɪkl] adj. 地理学的；地理的

infrastructure [ˈɪnfrəstrʌktʃər] n.基础下部组织；下部构造
navigation [nævɪˈgeɪʃn] n.导航；航海；航空；领航；航行
quarterly [ˈkwɔːtəli] adj.一年四次的；每季的
evolve [iˈvɒlv] v.（使）发展；（使）进展；（使）进化
cadastre [kəˈdɑːstreɪ] n.地籍簿；地籍；地籍图
cadastral surveying 地籍测量
sensor [ˈsensər] n.传感器
manipulate [məˈnɪpjuleɪt] vt.(熟练地)操作；使用(机器等)；操纵(人或市价、市场)；利用
state-of-the-art adj.先进的；一流的
geophysics [dʒiːəʊˈfɪzɪks] n.地球物理学
oceanography [əʊʃəˈnɒgrəfi] n.海洋学
retrieval [rɪˈtriːvl] n.检索；恢复；修补；重获
embrace [ɪmˈbreɪs] vt.拥抱；包含；收买；信奉 n.拥抱

☞ Notes：

① 测绘学（Geomatics）这个术语最初作为一门学科专业出现于加拿大，在过去几十年里已被世界各地众多的高等教育机构采纳，大多数是由以前的"大地测量学"或"测量学"并引入许多计算机科学和/或地理信息系统方向的课程后重新命名的。
② 采纳这一术语也使得在世界范围内,行业与学校之间在职场上取得了一致。
③ 加拿大测绘研究所在它们的季刊 *Geomatica* 中给出如下定义：测绘学（Geomatics）是运用系统的方法，集成各种手段来获取和管理空间数据，并作为科学、管理、法律和技术服务的一部分参与空间信息生产和管理的一门应用学科。

Terms Highlights

geomatics 测绘学
geodesy 大地测量学
surveying and mapping 测绘
photogrammetry 摄影测量学
remote sensing（RS）遥感
global positioning system（GPS）全球定位系统
satellite positioning 卫星定位
geographic information systems（GIS）地理信息系统
land management 土地管理
computer graphics 计算机图形学

Unit 2　Geodetic Surveying and Plane Surveying

Surveying has been traditionally defined as the art and science of determining the position of natural and artificial features on, above or below the Earth's surface; and representing this information in analog form as a contoured map, paper plan or chart, or as figures in report tables, or in digital form as a three dimensional mathematical model stored in the computer.① As such, the surveyor/geodesist dealt with the physical and mathematical aspect of measurement. The accurate determination and monumentation of points on the surface of the Earth is therefore seen as the major task. Though these surveys are for various purposes, still the basic operations are the same — they involve measurements and computations or, basically, fieldwork and office work. There are many different types of surveys such as land surveys, route surveys, city surveys, construction surveys, hydrographic surveys, etc., but generally speaking, surveying is divided into two major categories: geodetic and plane surveying. Surveys will either take into account the true shape of the Earth (geodetic surveys) or treat the Earth as a flat surface (plane surveys). Additionally, surveys are conducted for the purpose of positioning features on the ground (horizontal surveys), determining the elevation or heights of features (vertical surveys) or a combination of both.

Geodetic Surveying

The type of surveying that takes into account the true shape of the Earth is called *geodetic surveying*. This type of survey is suited for large areas and long lines and is used to find the precise location of basic points needed for establishing control for other surveys. In geodetic surveys, the stations are normally long distances apart, and more precise instruments and surveying methods are required for this type of surveying than for plane surveying.

Widely spaced, permanent monuments serve as the basis for computing lengths and distances between relative positions. These basic points with permanent monuments are called geodetic control survey points, which support the production of consistent and compatible data for surveying and mapping projects. In the past, ground-based theodolites, tapes, and electronic devices were the primary geodetic field measurements used. Today, the technological expansion of GPS has made it possible to perform extremely accurate geodetic surveys at a fraction of the cost. A thorough knowledge of the principles of geodesy is an absolute prerequisite for the proper planning and execution of geodetic surveys.

In geodetic surveys, the shape of the Earth is thought of as a spheroid, although in a technical sense, it is not really a spheroid. <u>Therefore, distances measured on or near the surface of the Earth are not along straight lines or planes, but on a curved surface. Hence, in the computation of distances in geodetic surveys, allowances are made for the Earth's minor and major diameters from which a spheroid of reference is developed.</u>[②] The position of each geodetic station is related to this spheroid. The positions are expressed as latitudes (angles north or south of the equator) and longitudes (angles east or west of a prime meridian) or as northings and eastings on a rectangular grid.

A geodetic survey establishes the fundamentals for the determination of the surface and gravity field of a country. This is realized by coordinates and gravity values of a sufficiently large number of control points, arranged in geodetic and gravimetric networks. In this fundamental work, curvature and the gravity field of the earth must be considered.

Plane Surveying

The type of surveying in which the mean surface of the Earth is considered a plane, or in which the curvature of the Earth can be disregarded without significant error, generally is called *plane surveying*. The term is used to designate survey work in which the distances or areas involved are of limited extent. With regard to horizontal distances and directions, a level line is considered mathematically straight, the direction of the plumb line is considered to be the same at all points within the limits of the survey, and all angles are considered to be plane angles.

To make computations in plane surveying, you will use formulas of plane trigonometry, algebra, and analytical geometry. For small areas, precise results may be obtained with plane surveying methods, but the accuracy and precision of such results will decrease as the area surveyed increases in size. <u>For example, the length of an arc 18.5km long lying in the Earth's surface is only 7mm greater than the subtended chord and, further, the difference between the sum of the angles in a plane triangle and the sum of those in a spherical triangle is only 0.51 second for a triangle at the Earth's surface having an area of 100km^2.</u>[③] It will be appreciated that the curvature of the Earth must be taken into consideration only in precise surveys of large areas.

A great number of surveys are of the plane surveying type. Surveys for the location and construction of highways, railroads, canals, and in general, the surveys necessary for the works of human beings are plane surveys, as are the surveys made to establish boundaries, except state and national. However, with the increasing size and sophistication of engineering and other scientific projects, surveyors who restrict their practice to plane surveying are severely limited in the types of surveys in which they can be engaged.

The operation of determining elevation usually is considered a division of plane surveying. Elevations are referred to the geoid. The geoid is theoretical only. It is the natural extension of the mean sea level surface under the landmass. We could illustrate this idea by digging an imaginary trench across the country linking the Atlantic and Pacific oceans. If we allowed the trench to fill with seawater, the surface of the water in the trench would represent the geoid. So for all intents and purposes, the geoid is the same as mean sea level. Mean sea level is the average level of the ocean surface halfway between the highest and lowest levels recorded. We use mean sea level as a datum or, curiously and incorrectly, a datum plane upon which we can reference or describe the heights of features on, above or below the ground.

Imagine a true plane tangent to the surface of mean sea level at a given point. At horizontal distances of 1km from the point of tangency, the vertical distances (or elevations) of the plane above the surface represented by mean sea level are 7.8cm. Obviously, curvature of the Earth's surface is a factor that cannot be neglected in obtaining even rough values of elevations. <u>The ordinary procedure in determining elevations, such as balancing backsight and foresight distance in differential leveling, automatically takes into account the curvature of the Earth and compensates for earth curvature and refraction, and elevations referred to the curved surface of reference are secured without extra effort by the surveyor.</u>④

There is a close cooperation between geodetic surveying and plane surveying. The geodetic survey adopts the parameters determined by measurements of the Earth, and its own results are available to those who measure the Earth. The plane surveys, in turn, are generally tied to the control points of the geodetic surveys and serve particularly in the development of national map series and in the formation of real estate cadastres.

Words and Expressions

artificial [ˌɑːtɪˈfɪʃl] *adj.* 人造的；假的；非原产地的
analog [ˈænəlɒg] *n.* 类似物；相似体
chart [tʃɑːt] *n.* 图表；海图
dimensional [daɪˈmenʃənl] *adj.* 空间的
monument [ˈmɒnjumənt] *n.* 纪念碑
permanent monument 永久标石
monumentation [ˌmɒnjumənˈteɪʃən] *n.* 埋石
fieldwork [ˈfiːldwɜːk] *n.* 野外工作；实地调查
category [ˈkætɪɡəri] *n.* 种类；类别；[逻]范畴
permanent [ˈpɜːmənənt] *adj.* 永久的；持久的
theodolite [θiˈɒdəlaɪt] *n.* [测]经纬仪
prerequisite [ˌpriːˈrekwəzɪt] *n.* 先决条件

spheroid [ˈsfɪrɔɪd] n.球状体；回转椭圆体
allowance [əˈlaʊəns] n.容许误差；容差；容许量
diameter [daɪˈæmɪtər] n.直径
equator [ɪˈkweɪtər] n.赤道；赤道线
latitude [ˈlætɪtjuːd] n.纬度；范围；(复数)地区
longitude [ˈlɒŋɡɪtjuːd] n.经度；经线经度
meridian [mɪˈrɪdiən] n.子午线；正午；顶点；全盛时期 adj.子午线的；正午的
prime meridian 本初子午线；本初子午圈线
northing [ˈnɔːθɪŋ] n.北距(向北航行的距离)；北进；北航
easting [ˈiːstɪŋ] n.东西距；朝东方；东行航程
gravity [ˈɡrævɪti] n.重力；地心引力
gravity field 重力场
curvature [ˈkɜːvətʃə(r)] n.曲率；弯曲
plumb [plʌm] n.铅锤；铅弹 adj.垂直的 vt.使垂直；探测
plumb line 铅垂线
trigonometry [trɪˈɡɒnəmi] n.三角法
plane trigonometry 平面三角
algebra [ˈældʒɪbrə] n.代数学
analytical [ˌænəˈlɪtɪkl] adj.解析的；分析的
analytical geometry 解析几何
chord [kɔːd] n.弦；弦长
triangle [ˈtraɪæŋɡl] n.[数]三角形；三人一组；三角关系
spherical [sferɪk(ə)l] adj.球形的；球的
sophisticate [səˈfɪstɪkeɪt] vt.弄复杂；篡改；使变得世故
sophistication [səˌfɪstɪˈkeɪʃn] n.复杂；强词夺理；诡辩
geoid [ˈdʒiːɔɪd] n.[地]大地水准面
trench [trentʃ] n.沟渠；堑壕；管沟；电缆沟；战壕
Atlantic Ocean 大西洋
Pacific Ocean 太平洋
tangent [ˈtændʒnt] adj.相切的；切线的 n.切线；[数]正切
backsight [ˈbæksaɪt] n.后视
foresight [ˈfɔːrsaɪt] n.前视；远见；深谋远虑
refraction [rɪˈfrækʃn] n.折光；折射

☞ **Notes**：

① 测量学的传统定义是：确定地球表面、地上及地下自然和人工建筑物位置，并通过等高线图、纸图和海图等模拟形式，或报表数据形式，或以三维数学模型存储在计算机中的数字形式来描绘这些信息的一门科学和技术。
② 因此，在地表或其附近所测得的距离并不是直线距离或平面上的距离，而是在曲面上

的距离。所以，在大地测量中计算距离时，应根据由地球长、短半轴所形成的参考椭球来设定容许误差。

③ 例如，地球表面一段为 18.5km 的弧段仅比它所对应的弦长多了 7mm，地球表面 100km² 的区域内平面三角形与球面三角形的内角和之差仅为 0.51″。

④ 测量高程的常规作业，如微差水准测量采用前、后视距平衡，则自动考虑了地球曲率，并补偿了地球曲率和大气折光的影响，不需要测量员的额外努力就可以准确获得相对于基准面的高程。

Terms Highlights

geodetic surveying 大地测量；大地测量学
plane surveying 平面测量；平面测量学
control survey 控制测量
horizontal survey 水平测量；平面测量
vertical survey 高程测量；垂直测量
topographic survey 地形测量
detail survey 碎部测量
land survey (property survey, boundary survey, cadastral survey) 土地测量；地籍测量
route survey 路线测量
pipe survey 管道测量
city survey 城市测量
hydrographic survey 水道测量
marine survey 海洋测量
mine survey 矿山测量
geological survey 地质测量

Unit 3 Distance Measurement

One of the fundamentals of surveying is the need to measure distance. Distances are not necessarily linear, especially if they occur on the spherical Earth. In this subject we will deal with distances in Euclidean space, which we can consider a straight line from one point or feature to another. Distance between two points can be horizontal, slope, or vertical. Horizontal and slope distances can be measured with lots of techniques of measurement depending on the desired quality of the result. In plane surveying, the distance between two points means the horizontal distance. If the points are at different elevations, then the distance is the horizontal length between plumb lines at the points. Here gives a brief summary of relevant techniques and their respective accuracies:

Pacing and Odometer

Pacing is a very useful form of measurement though it is not precise, especially when surveyors are looking for survey marks in the field. Pacing can be performed at an accuracy level of 1/100~1/500 when performed on horizontal land, while the accuracy of pacing can't be relied upon when pacing up or down steep hills. The odometer is a simple device that can be attached to any vehicle and directly registers the number of revolutions of a wheel. With the circumference of the wheel known, the relation between revolutions and distance is fixed.

Ordinary Taping and Precise Taping

Taping is a very common technique for measuring horizontal distance between two points. Ordinary taping refers to the very common tapes that we can buy them in stores, such as the plastic tapes or poly tapes. Such tapes have low precision in distance measurements with about 1/3000~1/5000. The precise taping refers to the steel tapes and which are much more expensive than the plastic tape and have higher precision of 1/10000~1/30000. Invar tapes are composed of 35% nickel and 65% steel. This alloy has a very low coefficient of thermal expansion, making the tapes useful in precise distance measurement. Many tapes are now graduated with foot units on one side and metric units on the reverse side. Metric units are in meters, centimeter and minimeter with the total length of 20m, 30m, 50m and 100m.

If we want to measure the horizontal distance between the two points A and B, we can do like this: With zero of the tape to the higher point B and tape going along the point A, we can measure the horizontal distance by using the plumb bob with pump line centering to the point A. To judge the exact horizontal line, we should move the tape up and down along the pump line and we will find the changes of reading in the tape. The shortest reading of the tape is the horizontal distance. If the distance is longer than the length of tape, then we can divide the long distance into several segments and get the total distance by plus each segment together. Since different tapes have different starts of zero of the tapes, it is very important to judge where the zero of the tape begins.

Tacheometry and Stadia

Tacheometry is an optical solution to the measurement of distance. The word is derived from the Greek Tacns, meaning "swift", and metrot, meaning "a measure". Tacheometry involves the measurement of a related distance parameter either by means of a fixed-angle intercept. Theodolite tacheometry is an example of stadia system. The theodolite is directed at the level staff where the staff is held vertically and the line of sight of the telescope is horizontal. By reading the top and bottom stadia hairs on the telescope view and then the horizontal distance from center of instrument to rod can be obtained by multiplying the stadia interval factor K by the stadia interval and plus the distance C which is from the center of instrument to principal focus, i.e., $D = Ks + C$.[①] Usually the nominal stadia interval factor K equals 100 which is a constant for a particular instrument as long as conditions remain unchanged, but it may be determined by observation in practice. The value of C is determined by the manufacturer and stated on the inside of the instrument box. For external-focusing telescopes, under ordinary condition, C may be considered as 1 ft without error of consequence. Internal-focusing telescopes are so constructed that C is 0 or nearly so; this is an advantage of internal-focus telescopes for stadia work. Most instruments now used for stadia are equipped with internal-focusing telescopes.

Applications of tacheometry include traversing and leveling for the topographic surveys, location of detail surveys, leveling and field completion surveys for the topographic mapping, and hydrographic mapping. The relative precision is 1 : 1000 to 1 : 5000.

Stadia is a form of tacheometry that uses a telescopic cross-hair configuration to assist in determining distances. A series of rod readings is taken with a theodolite and the resultant intervals are used to determine distances.

Electronic Distance Measurement (EDM)

The Electronic Distance Measurement (EDM) was first introduced in 1950s by the founders of Geodimeter Inc. The advent of EDM instrument has completely revolutionized all surveying procedures, resulting in a change of emphasis and techniques. Distance can now be measured easily, quickly and with great accuracy, regardless of terrain conditions.

EDM instruments refer to the distance measurement equipments using light and radio waves. Both light waves and radio waves are electromagnetic. They have identical velocities in a vacuum (or space) to 299,792.458±0.001km/sec. These velocities, which are affected by the air's density, are reduced and need to be recalculated in the atmosphere. The basic principle of EDM instruments is that distance equals time multiplied by velocity. Thus if the velocity of a radio or light wave and time required for it to go from one point to another are known, the distance between the two points can be calculated.

The EDM instruments may be classified according to the type and wavelength of the electromagnetic energy generated or according to their operational range. EDM instruments use three different wavelength bands: (1) *Microwave systems* with range up to 150km, wavelength 3cm, not limited to line of sight and unaffected by visibility; (2) *Light wave systems* with range up to 5km (for small machines), visible light, lasers and distance reduced by visibility; (3) *Infrared systems* with range up to 3km, limited to line of sight and limited by rain, fog, other airborne particles.

Although there is a wide variety of EDM instruments available with different wavelengths, there are basically only two methods of measurement employed which may divide the instruments into two classification as electro-optical (light waves) and microwaves (radio waves) instruments.② These two basic methods are namely the pulse method and more popular phase different method. They function by sending light waves or microwaves along the path to be measured and measuring the time differences between transmitted and received signals, or measuring the phase differences between transmitted and received signals in returning the reflecting light wave to source.③

Modern EDM instruments are fully automatic to such an extent that, after the instruments, set up on one station, emits a modulated light beam to a passive reflector set up on the other end of the line to be measured. The operator need only depress a button, and the slope distance is automatically displayed.④ More complete EDM instruments also have the capability of measuring horizontal and vertical or zenith angles as well as the slope distance. These instruments referred to as total station instruments.

Words and Expressions

fundamental [fʌnˈdæmentl] *n.* 基本原则；基本原理
Euclidean space 欧几里得空间
odometer [əˈdɒmɪtər] *n.* (汽车等的)里程表；自动计程仪
vehicle [ˈviːəkl] *n.* 交通工具；车辆；媒介物；传达手段
revolution [ˌrevəˈluːʃn] *n.* 旋转；革命

circumference [sɜːrˈkʌmfərəns] n.圆周；周围
invar [ˈɪnvɑːr] n.因瓦；不胀钢
nickel [ˈnɪkl] n.[化]镍；镍币；(美国和加拿大的)五分镍币
alloy [ˈælɔɪ] n.合金
coefficient [kəʊˈfɪʃnt] n.[数]系数
thermal [ˈθɜːməl] adj.热的；热量的
tacheometry [tækɪˈɒmɪtrɪ] n.[测]视距测量
stadia [ˈsteɪdɪə] n.视距；视距仪器
intercept [ˌɪntəˈsept] vt.截取；中途阻止
telescope [ˈtelɪskəʊp] n.望远镜
multiply [ˈmʌltɪplaɪ] v.乘；增加；繁殖
nominal [ˈnɒmɪnl] adj.名义上的；有名无实的；名字的；[语]名词性的
manufacturer [ˌmænjuˈfæktʃərə(r)] n.制造业者；厂商
consequence [ˈkɒnsɪkwəns] n.结果 [逻]推理；推论；因果关系；重要的地位
topographic [tɒpəˈɡræfɪk] adj.地势的；地形学上的
resultant [rɪˈzʌltənt] adj.作为结果而发生的；合成的
terrain [təˈreɪn] n.地形
electromagnetic [ɪˌlektrəʊmæɡˈnetɪk] adj.电磁的
visibility [ˌvɪzəˈbɪləti] n.可见度；能见度；可见性；显著；明显度
infrared [ˌɪnfrəˈred] adj.红外线的 n.红外线
airborne [ˈeəbɔːn] adj.空气传播的；空降的；空运的
particle [ˈpɑːtɪkl] n.粒子；点；极小量；微粒；质点
modulated [ˈmɒdjʊleɪtɪd] adj.已调整[制]的；被调的

☞ Notes：

① 在望远镜视窗中读取上下丝的读数，则可以得到仪器中心到标尺的水平距离为视距间隔因子 K 与视距间隔相乘再加上经纬仪中心到主焦点的距离 C，即为 $D=Ks+C$。
② 尽管电子测距仪种类很多，采用的波长各异，但所用的测量方法基本上可以分为两种，从而将测距仪划分为两类：光电(光波)测距仪和微波(无线电波)测距仪。
③ (工作原理如下：)这两种测距仪都是通过沿着待测路径发射光波和微波，测得发射和接收信号的时间差，或者测量发射和接收反射光返回光源处的信号相位差来进行距离测量。
④ 现代电子测距仪(EDM)的全自动化已达到这样的程度：当在测站架好仪器后，测距仪向架设在测线另一端的被动反射镜上发射调制光波，操作员只需要按一下按钮，斜距就可以自动显示出来。

Terms Highlights

distance measurement 距离测量

precise ranging 精密测距
pacing 步测；定步
distance measuring instrument, rangefinder 测距仪
EDM(electronic distance measurement) 电子测距仪
geodimeter 光速测距仪；光电测距仪
electromagnetic distance measuring instrument 电磁波测距仪
electro-optical distance measuring instrument 光电测距仪
long-range EDM instrument 远程电子测距仪
infrared EDM instrument 红外测距仪
laser distance measuring instrument, laser ranger 激光测距仪
microwave distance measuring instrument 微波测距仪
satellite laser ranger 卫星激光测距仪
two-color laser ranger 双色激光测距仪
distance-measuring error 测距误差
fixed error 固定误差
proportional error 比例误差
sighting distance 视距
multiplication constant 乘常数
addition constant 加常数
stadia multiplication constant 视距乘常数
stadia addition constant 视距加常数
standard field of length 长度标准检定场
nominal accuracy 标称精度
stadia hair 视距丝；视距线
stadia interval 视距间隔

Unit 4 Angle and Direction Measurement

Horizontal and vertical angles are fundamental measurements in surveying. It is necessary to be familiar with the meanings of certain basic terms before describing angle and direction measurement. The terms discussed here have reference to the actual figure of the Earth.

Basic Terms

A *vertical line* at any point on the Earth's surface is the line that follows the direction of gravity at that point. It is the direction that a string will assume if a weight is attached at that point and the string is suspended freely at the point. At a given point there is only one vertical line.

A *horizontal line* at a point is any line that is perpendicular to the vertical line at the point. At any point there are an unlimited number of horizontal lines.

A *horizontal plane* at a point is the plane that is perpendicular to the vertical line at the point. There is only one horizontal plane through a given point.

A *vertical plane* at a point is any plane that contains the vertical line at the point. There are an unlimited number of vertical planes at a given point.

Horizontal Angle and Vertical Angle

A horizontal angle is the angle formed in a horizontal plane by two intersecting vertical planes, or a horizontal angle between two lines is the angle between the projections of the lines onto a horizontal plane. For example, observations to different elevation points B and C from A will give the horizontal angle $\angle bac$ which is the angle between the projections of two lines (AB and AC) onto the horizontal plane. It follows that, although the points observed are at different elevations, it is always the horizontal angle and not the space angle that is measured (Figure 1). The horizontal angle is used

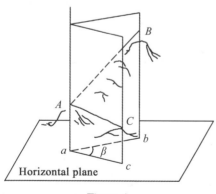

Figure 1

primarily to obtain relative direction to a survey control point, or topographic detail points, or to points to be set out.

A vertical angle is an angle measured in a vertical plane which is referenced to a horizontal line by plus (up) or minus (down) angles, or to a vertical line from the zenith direction. Plus and minus vertical angles are sometimes referred to as elevation or depression angles, respectively. A vertical angle thus lies between 0° and ±90°. *Zenith* is the term describing points on a celestial sphere that is a sphere of infinitely large radius with its center at the center of the Earth. The zenith is an angle measured in a vertical plane downward from an upward directed vertical line through the instrument. It is thus between 0° and 180°. Obviously the zenith angle is equal to 90° minus the vertical angles. Vertical angles or zeniths are used in the correction of slope distance to the horizontal or in height determined.

For the most part, the instrument used in the measurement of angles is called a transit or theodolite, although angles can be measured with clinometers, sextants (hydrographic surveys), or compasses. The theodolite contains a horizontal and vertical circles of either glass or silver. The horizontal and vertical circles of theodolite can be linked to circular protractors graduated from 0° to 360° in a clockwise manner set in horizontal and vertical plane.[①] The horizontal circle is used when measuring or laying off horizontal angles and the vertical circle is used to measure or lay off vertical angles or zenith angles. Usually the units of angular measurement employed in practice are degrees, minutes, and seconds, the sexagesimal system.

Angle Measurement

A horizontal angle in surveying has a direction or sense; that is, it is measured or designed to the right or to the left, or it is considered clockwise or counterclockwise. In the above figure, the angle at A from B to C is clockwise and the angle from C to B is counterclockwise. With the theodolite set up, centered, and leveled over at station A, then a simple horizontal angle measurement between surveying points B, A and C would be taken as follows:

(1) Commencing on, say, "face left", the target set at survey point B is carefully bisected and the reading on horizontal scale is 25°.

(2) The upper plate clamp is released and telescope is turned clockwise to survey point C. The reading on horizontal circle is 75°.

(3) The horizontal angle is then the difference of the two directions, i.e. (75°−25°) = 50°.

(4) Change face and observe point C on "face right", and note the reading=255°.

(5) Release upper plate and swing counterclockwise to point B and note the reading =205°.

(6) The reading or the direction must be subtracted in the same order as 255°−205°

=50°.

(7) The mean of two values would be accepted if they are in acceptable agreement.

Modern electronic digital theodolites contain circular encoders that sense the rotations of the spindles and the telescope, convert these rotations into horizontal and vertical (or zenith) angles electronically, and display the value of the angles on liquid crystal displays (LCDs) or light-emitting diode displays (LEDs). These readouts can be recorded in a conventional field book or can be stored in a data collector for future printout or computation. The instrument contains a pendulum compensator or some other provision for indexing the vertical circle readings to an absolute vertical direction.② The circle can be set to zero readings by a simple press of a button or initialized to any value on the instrument.

Azimuth is the horizontal angle measured in a clockwise direction from the plane of the meridian, which is a line on the mean surface of the Earth joining the north and south poles. Azimuth ranges in magnitude from 0° to 360°, values in excess of 360°, which are sometimes encountered in computations, are simply reduced by 360° before final listing.③

Bearing is the traditional way of stating the orientation of the line. It is actually the angle measured from the north or south. The bearing, which can be measured clockwise or counterclockwise from the north or south end of the meridian, is always accompanied by letters that locate the quadrant in which the line falls.④ For example, bearing N32°W indicates a line trending 32° west of the north. It is equal to an azimuth of 328°. Bearing S12°W indicates a line trending 12° west of the south. It is equal to an azimuth of 192°.

It is important to state that the bearing and azimuth are respect to true north.

Words and Expressions

perpendicular [ˌpɜːpenˈdɪkjələ(r)] *adj.* 垂直的；正交的
intersect [ɪntəˈsekt] *vt.* 横断 *vi.* (直线)相交；交叉
projection [prəˈdʒekʃn] *n.* 投影；投射；投影图；地图投影；规划
zenith [ˈzenɪθ] *n.* 天顶；顶点；顶峰；最高点
celestial [ˈsɛlestʃəl] *adj.* 天上的
celestial sphere 天球
radius [ˈreɪdiəs] *n.* 半径；范围；辐射光线；有效航程；界线
clinometer [klɪˈnɒmɪtər] *n.* 测角器；倾斜仪
sextant [ˈsekstənt] *n.* 六分仪
compass [ˈkʌmpəs] *n.* 罗盘；指南针；圆规
protractor [prəˈtræktə(r)] *n.* 量角器
clockwise [ˈklɒkwaɪz] *adj.* 顺时针方向的
counterclockwise [ˌkaʊntərˈklɒkwaɪz] *adj.* 反时针方向的
sexagesimal [seksəˈdʒɪsm(ə)l] *adj.* 六十的；六十进位的

sexagesimal system 六十分制
commence [kəˈmens] v.开始；着手
bisect [baɪˈsekt] v.切成两份；对(截)开
clamp [klæmp] n.夹子；夹具；夹钳
encoder [ɪnˈkəʊdər] n.编码器；译码器
spindle [ˈspɪndl] n.轴；杆；心轴；锭子；纺锤
crystal [ˈkrɪstl] adj.结晶状的 n.水晶；水晶饰品；结晶；晶体
liquid crystal displays (LCDs) 液晶显示
diode [ˈdaɪəʊd] n.二极管
light-emitting diode displays (LEDs) 发光二极管显示
pendulum [ˈpendʒələm] n.钟摆；摇锤
compensator [ˈkɒmpenseɪtə] n.补偿器
provision [prəˈvɪʒn] n.供应；(一批)供应品；预备；防备；规定
indexing [ˈɪndeksɪŋ] n.标定指数
initialize [ɪˈnɪʃəlaɪz] vt.初始化
azimuth [ˈæzɪməθ] n.方位；方位角
bearing [ˈbeərɪŋ] n.方向；方位
quadrant [ˈkwɒdrənt] n.象限；四分仪

☞ Notes:

① 经纬仪的水平度盘和垂直度盘可以与水平和垂直放置的、按顺时针方向从0°到360°均匀刻划的圆分度器相连。
② 电子经纬仪包含一个钟摆补偿器或其他装置，可以得到绝对竖直方向的竖盘指标读数。
③ 方位角的取值范围为0°~360°。计算中可能会遇到方位角值超过360°的情况，最终结果减去360°即可。
④ 方向角可以从子午线的北端或南端顺时针或逆时针测得，表示时通常加上该方向角所在象限对应的字母。

Terms Highlights

horizontal angle 水平角
vertical angle 垂直角
depression angle 俯角；俯视角
zenith distance 天顶距
elevation angle 高度角
horizontal circle 水平刻度盘
vertical circle 垂直度盘
true north 真北
geodetic azimuth 大地方位角

grid bearing 坐标方位角
gyro azimuth 陀螺方位角
magnetic azimuth 磁方位角
method by series, method of direction observation 方向观测法
method in all combinations 全组合测角法

Unit 5 Traversing

The purpose of the surveying is to locate the positions of points on or near the surface of the Earth. To determine horizontal positions of arbitrary points on the Earth's surface and elevation of points above or below a reference surface are known as a control survey. The positions and elevations of the points make up a control network. There are different types of control networks depending on where and why they are established. A control network may have very accurate positions but no elevations (called a Horizontal Control Network) or very accurate elevations but no positions (called a Vertical Control Network). Some points in a control network have both accurate positions and elevations. Control networks range from small, simple and inexpensive to large and complex and very expensive to establish.

<u>A control network may cover a small area by using a "local" coordinate system that allows you to position the features in relation to the control network but doesn't tell you where the features are on the surface of the Earth, or cover a large area by consisting of a few well-placed and precise-established control points, which is sometimes called the primary control.</u> [1] The horizontal positions of points in a network can be obtained in a number of different ways. The generally used methods are triangulation, trilateration, traversing, intersection, resection and GPS. The main topic of this text refers to the traversing.

Triangulation

The method of surveying called triangulation is based on the trigonometric proposition that if one side and three angles of a triangle are known, the remaining sides can be computed by the law of sines. Furthermore, if the direction of one side is known, the direction of the remaining sides can be determined. And then coordinates of unknown points can be computed by application of trigonometry.

Trilateration

Since the advent of long-range EDM instrument, a method of surveying called trilateration was adopted to combine with triangulation. The trilateration is based on the trigonometric proposition that if the three sides of a triangle are known, the three angles can be computed by the law of cosines. Trilateration possesses some advantages over triangulation because the measurement of the distances with EDM instrument is so quick, precise and economical while the measurement of the angles needed for triangulation may be more difficult and expensive. For some precise projects, the combination of triangulation and trilateration which is called triangulateration is applied.

Traversing

A survey traverse is a sequence of lengths and directions of lines between points on the Earth, obtained by or from field angle and distance measurements and used in determining positions of the points. The angles are measured using transits, theodolites, or total stations, whereas the distances can be measured using steel tapes or EDM instruments. A survey traverse may determine the relative positions of the points that it connects in series, and if tied to control stations based on some coordinate system, the positions may be referred to that system. From these computed relative positions, additional data can be measured for layout of new features, such as buildings and roads.

Since the advent of EDM equipment, traversing has emerged as the most popular method to establish control networks such as basic area control, mapping, control of hydrographic surveys and construction projects. In engineering surveying, it is ideal way to surveys and dimensional control of route-type projects such as highway, railroad, and pipeline construction. In general, a traverse is always classified as either an *open traverse* or a *closed traverse*. An open traverse originates either at a point of known horizontal position with respect to a horizontal datum or at an assumed horizontal position, and terminates at a station whose relative position is not previously known.[②] The open traverse provides no check against mistakes and large errors for its termination at an unknown horizontal position and lack of geometric closure. This lack of geometric closure means that there is no geometric verification possible with respect to the actual positioning of the traverse stations. Thus, the measuring technique must be refined to provide for field verification. At a minimum, distances are measured twice and angles are doubled. Open traverses are often used for preliminary survey for a road or railroad. A closed traverse can be described in any one of the following two ways: (1) A closed loop traverse, as the name implies, forms a continuous loop, enclosing an area. This type of closed traverse starts at assumed horizontal position or at a known horizontal position with respect to a horizontal datum and ends at the same point. (2) A connecting traverse starts and ends at separate points, whose relative positions have been determined by a survey of equal or higher order accuracy. A known horizontal position is defined by its geographic latitude and longitude, or by its X and Y coordinates on a grid system. Closed traverses, whether they return to the starting point or not, provide checks on the measured angles and distances. In both cases, the angles can be closed geometrically, and the position closure can be determined mathematically. Therefore they are more desirable and used extensively in control, construction, property, and topographic surveys.

As we mentioned above, a closed traverse provides checks on the measured angles and distances. For example, the geometric sum of the interior angles in an n-side closed figure should be $(n-2) \cdot 180°$, but due to systematic and random errors of the measurements, when all the interior angles of a closed traverse are summed, they may or may not total the number of degrees required for geometric closure. The difference between the geometric sum and actual field sum of the interior angles is called angular closure. The total error of angular closure should be distributed evenly to each angle (if all angles were measured with the same precision) before mathematical analysis of the traverse. The important point before doing this is that the overall angular closure can't be beyond the survey specifications.

<u>Closed traverses provide also checks on the measured distances, and the position closure can be determined mathematically, which means an indication of the consistency of measuring distances as well as angles should be given to a traverse that closes on itself.</u>[3] Theoretically this position closure from the origin back to itself should be zero. <u>But the errors in the measured distances and angles of the traverses, however, will tend to alter the shape of the traverse, therefore we should compute the algebraic sum of the latitudes and algebraic sum of the departures, and compare them with the fixed latitude and departure of a straight line from the origin to the closing point.</u>[4] By definition, latitude here is the north/south rectangular component of a line and departure is the east/west rectangular component of a line. To differentiate direction, north is considered plus, whereas south is considered minus. Similarly, east is considered plus, whereas west is considered minus. Then the discrepancy should be adjusted by apportioning the closure both in latitudes and in departures on a reasonable basis. The adjusted position of each traverse point is determined with respect to some origin. This position is defined by its Y coordinate and its X coordinates with respect to a plane rectangular coordinate system in which the Y axis is assumed north-south whereas the X axis east-west.

Words and Expressions

traverse [træˈvɜːs] n. 导线；横贯；横断
traversing [træˈvɜːsɪŋ] n. 导线测量
arbitrary [ˈɑːrbɪtreri] adj. 任意的；武断的；独裁的；专断的
triangulation [traɪˌæŋɡjʊˈleɪʃn] n. 三角测量
trilateration [ˌtraɪlætəˈreɪʃn] n. 三边测量
triangulateration [ˌtraɪæŋɡjʊˈleɪtəreɪʃn] n. 边角测量
proposition [ˌprɒpəˈzɪʃn] n. 命题；主张；建议；陈述
law of sines 正弦定律
law of cosine 余弦定律
terminate [ˈtɜːrmɪneɪt] v. 停止；结束；终止

preliminary [prɪˈlɪmɪneri] *adj.* 初步的；预备的
property [ˈprɒpəti] *n.* 所有物；所有权；性质；特性；(小)道具
evenly [ˈiːvnli] *adv.* 均匀地；平坦地
indication [ˌɪndɪˈkeɪʃn] *n.* 指出；指示；迹象；暗示
consistency [kənˈsɪstənsi] *n.* 一致性；连贯性；结合；坚固性
alter [ˈɔːltər] *v.* 改变
departure [dɪˈpɑːrtʃər] *n.* 横距；偏移
rectangular [rɪkˈtæŋɡjʊlər] *adj.* 矩形的；成直角的
discrepancy [dɪsˈkrepənsi] *n.* 差异；相差；矛盾
apportion [əˈpɔːrʃn] *v.* 分配

☞ **Notes：**

① 小范围控制网可使用"局部"坐标系，局部坐标系能确定特征点相对控制网的位置，但不能给出这些特征点在地球表面上的什么位置；控制网也可以利用一些精确埋设和建立的控制点来覆盖大的区域，这种控制网有时称为首级控制。
② 支导线起始于某一已知的平面基准点或假定基准点，终止于一个相对位置事先未知的点。
③ 闭合导线也为距离测量提供了检核，其点位闭合差可以通过数学计算得到，这就意味着点位闭合差是指闭合导线的测距和测角应具备自身闭合的一致性。
④ 但由于导线距离和角度测量的误差可能会改变导线的形状，因此，我们应当计算(南北)纵距和(东西)横距的代数和，并将它们与已知的起点和终点连线的纵距和横距进行比较。

Terms Highlights

control network 控制网
horizontal control network 平面控制网；水平控制网
vertical control network 高程控制网
control point 控制点
triangulation 三角测量
trilateration 三边测量
triangulateration 边角测量
forward intersection 前方交会
resection 后方交会
side intersection 侧方交会
linear-angular intersection 边角交会法
linear intersection 边交会法
traversing 导线测量
traverse point 导线点

traverse leg 导线边
traverse angle 导线折角
open traverse 支导线
closed traverse 闭合导线
closed loop traverse 闭合环导线
connecting traverse 附合导线
angle closing error of traverse 导线角度闭合差
total length closing error of traverse 导线全长闭合差
closing error in coordinate increment 坐标增量闭合差
traverse network 导线网
triangulateration network 边角网
triangulation network 三角网
trilateration network 三边网
survey specifications, specifications of surveys 测量规范

Unit 6 Methods of Elevation Determination

An *elevation* is a vertical distance above or below a reference datum. Although vertical distance can be referenced to any datum, in surveying, the reference datum that is universally employed is that of *mean sea level* (*MSL*). MSL is assigned a vertical value (elevation) of 0.000*ft* or 0.000m. All other points on the Earth can be described by the elevations above or below zero. Permanent points whose elevations have been precisely determined (benchmarks) are available in most areas for survey use. In China, 7 years of observations at tidal stations in Qingdao from 1950 to 1956 were reduced and adjusted to provide the Huanghai vertical datum of 1956. In 1987, this datum was further refined to reflect long periodical ocean tide change to provide a new national vertical datum of 1985, according to the observations at tidal stations from 1952 to 1979.[①] Although, strictly speaking, the national vertical datum may not precisely agree with the MSL at specific points on the Earth's surface, the term MSL is generally used to describe the datum. MSL is assigned a vertical value (elevation) of 0.000*ft* or 0.000m.

Difference in elevation may be measured by the following methods: (James M. Anderson and Edward M.Mikhail, 1998)

1. *Direct* or *spirit leveling*, by measuring vertical distances directly. Direct leveling is most precise method of determining elevations and the one commonly used.
2. *Indirect* or *trigonometric leveling*, by measuring vertical angles and horizontal or slope distances.
3. *Stadia leveling*, in which vertical distances are determined by tacheometry using engineer's transit and level rod; plane-table and alidade and level rod; or self-reducing tacheometer and level rod.
4. *Barometric leveling*, by measuring the differences in atmospheric pressure at various stations by means of a barometer.
5. *Gravimetric leveling*, by measuring the differences in gravity at various stations by means of a gravimeter for geodetic purposes.
6. *Inertial positioning system*, in which an inertial platform has three mutually perpendicular axes, one of which is "up", so that the system yields elevation as one of the outputs.[②] Vertical accuracies from 15 to 50cm in distances of 60 and 100km, respectively, have been reported. The equipment cost is extremely high and applications are restricted to very large projects where terrain, weather, time, and access impose special constraints on traditional methods.

7. *GPS survey elevations* are referenced to the ellipsoid but can be corrected to the datum if a sufficient number of points with datum elevations are located in the region surveyed. Standard deviations in elevation differences of 0.053 to 0.094m are possible under these conditions.

Spirit Leveling

The most precise method of determining elevations and most commonly used method is direct leveling or spirit leveling which means measuring the vertical distance directly. Differential leveling is used to determine differences in elevation between points that are remote from each other by using a surveyor's level together with a graduated measuring rod. For example, to determine the elevations of desired point B with respect to a point of known elevation A (see Figure 1), the elevation of which (BM) is known to be H_A above sea level, the level is set up at intermediate point between A and B, and rod readings are taken at both locations as a and b respectively. Then the elevation of the line of sight of the instrument (being horizontal) is known to be the line of sight of the instrument $H_A + a$. The elevation of point B can be determined by equation

$$H_B = H_A + a - b$$

In addition to determining the elevation of point B, the elevations of any other points, lower than the line of sight and visible from the level, can be determined in a similar manner. But some terms should be mentioned from above. a is called Backsight (BS) which is a rod reading taken on a point of known elevation in order to establish the elevation of the instrument line of sight. b is called Foresight (FS) which is a rod reading

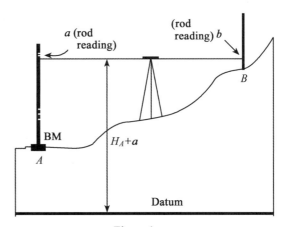

Figure 1

taken on a turning point, benchmark, or temporary benchmark in order to determine its elevation. $H_A + a$ refers to the Height of Instrument (HI) which is the elevation of the line of sight through the level.

Owing to refraction, actually the line of sight is slightly curved, the effects of curvature and refraction for the horizontal distance can be reduced to a negligible amount and no correction for curvature and refraction is necessary if backsight and foresight distances are balanced in practical operation.[3]

Trigonometric Leveling

Trigonometric leveling is used where difficult terrain, such as mountainous areas, precludes the use of conventional differential leveling.

The modern approach is to measure the slope distance and vertical angle to the point in question. Slope distance is measured using electromagnetic distance measurers and the vertical (or zenith) angle using a theodolite, or the total station that integrate these two instruments into a single instrument. Total stations contain built-in microprocessors that calculate and display the horizontal distance from the measured slope distance and vertical height. This latter facility has resulted in trigonometrical leveling being used for a wide variety of heighting procedures, including contouring. The basic concept of trigonometrical leveling can be seen from Figure 2. When measuring the vertical angle α and the horizontal distance S is used, then the difference in elevation h_{AB} between ground points A and B is therefore:

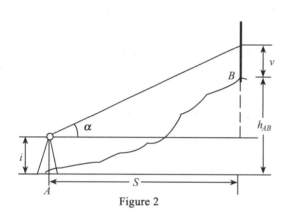

Figure 2

$$h_{AB} = S \cdot \tan\alpha + i - v$$

where i is the vertical height of the measuring center of the instrument above A and v is the vertical height of the center of the target above B. The vertical angles are positive for angles of elevation and negative for angles of depression. The zenith angles are always positive, but naturally when greater than 90° they will produce a negative result.

Trigonometrical leveling method of determining difference in elevation is limited to horizontal distance less than 300m when moderate precision is sufficient, and to proportionately shorter distances as high precision is desired. For the distance beyond 300m the effects of curvature and refraction must be considered and applied. To eliminate the uncertainty in the curvature and refraction correction, vertical-angle observations are made at both ends of the line as close in point of time as possible. This pair of observations is termed reciprocal vertical-angle observation. The correct difference in elevation between the two ends of the line is the mean of the two values computed both ways either with or without taking into account curvature and refraction.

The important notes should be mentioned here is that surveyors used to working with spirit levels have referenced orthometric heights (H) to the "average" surface of the Earth, as depicted by MSL.[④] However, the elevation coordinate (h) given by GPS solutions refers to the height from the surface of the ellipsoid to the ground station.

Words and Expressions

alidade [ˌalɪˈdeɪd] n.[测]照准仪
barometer [bəˈrɒmɪtə] n.气压计
inertial [ɪˈnɜːʃl] adj.惯性的；不活泼的
ellipsoid [ɪˈlɪpsɔɪd] n.椭圆体
deviation [ˌdiːviˈeɪʃn] n.偏差；偏移
intermediate [ˌɪntəˈmiːdiət] adj.中间的
preclude [prɪˈkluːd] n.排除
microprocessor [ˌmaɪkrəʊˈprəʊsesər] n.[计]微处理器
reciprocal [rɪˈsɪprəkl] adj.彼此相反的；互惠的；相应的；倒数的
orthometric [ɔːθəˈmetrɪk] adj.正高的
depict [dɪˈpɪkt] vt.描述；描写

☞ Notes：

① 根据验潮站 1952 年至 1979 年的观测数据，1987 年该基准得到进一步精化以反映海潮的长周期变化，新基准即为 1985 年国家高程基准。
② 惯性定位系统：该系统中的惯性平台有三个互相正交的轴，其中的一个轴向上。这样高程是该系统的输出结果之一。
③ 由于大气折光，实际上视线会有轻微的弯曲，在实际测量中如果保持前后视距相等，那么地球曲率和折光差对水平距离的影响可以减小到忽略不计的程度，不需要对曲率和折光差进行改正。
④ 这里应该重点提到的是，测量员施测几何水准测量是以地球的"平均"表面，即平均海水面所描述的正高(H)为参考面。

Terms Highlights

reference datum 参考基面；参考基准面
Huanghai vertical datum of 1956 1956 黄海高程系统
national vertical datum of 1985 1985 国家高程基准
direct leveling, spirit leveling 几何水准测量
differential leveling 微差水准测量
trigonometric leveling 三角高程测量
barometric leveling 气压水准测量
gravimetric leveling 重力水准测量
river-crossing leveling 跨河水准测量
BM(benchmark) 水准基点
level rod 水准尺

level 水准仪
backsight (BS) 后尺
foresight (FS) 前尺
height of instrument (HI) 仪器高
height of target (HT) 目标高
elevation difference 高差
annexed leveling line 附合水准路线
closed leveling line 闭合水准路线
spur leveling line 支水准路线
refraction correction 折光差改正
elevation of sight 视线高程
optical level 光学水准仪
electronic level 电子水准仪
automatic level, compensator level 自动安平水准仪
laser level 激光水准仪

Unit 7 Robotic Total Station

For many years, the optical transit was the surveyor's tool of choice to measure angles. By the 1970s, however, the electronic theodolite began to replace the transit since it could measure angles more accurately on both the horizontal and vertical axes. In the early 1980s, "total stations," which measure distances very accurately by using electronic distance meters (EDMs), became the instrument of choice. Then in late 1990, Geodimeter, Dandryd Sweden introduced the first "robotic total station" adding automatic tracking and radio communication to a radio and data collector at the "target" or pole. Thus, for the first time, no person was required at the instrument—only at the target, reducing the size of a survey crew.

Total Station

A total station is the most commonly used instrument now in geomatics engineering, which is fully integrated instrument that captures all the spatial data necessary for a 3-dimensional positional information. A total station integrates the functions of an electronic theodolite for measuring angles, an EDM for measuring distances, digital data and a data recorder. All total stations have similar constructional features regardless of their age or level of technology, and all perform basically the same functions.

After the instrument has been set up on a control station, centered, leveled and properly oriented, and the prism target has been set up over another point whose position is to be measured, the surveyor may focus the target and depress a button. Then output from the horizontal and vertical circular encoders and from the EDM can be displayed at the instrument and stored in a data collector and enters into a built-in microprocessor. The microprocessor can convert the measured slope distance to the horizontal distance using the measured vertical or zenith angle. The microprocessor also computes the difference in elevation between the instrument center and the prism target. If the elevation of the instrument center (the HI) and the height of the reflector target (the HT) above the ground are entered, the microprocessor computes the elevation of the target station taking into account the effect of curvature and refraction. Furthermore the microprocessor can also compute the resolution of the horizontal distance together with the current horizontal direction, expressed as an azimuth, into a coordinate of the target station. In construction layout measurement, the data necessary to establish the direction and distance from a control point to locate a construction point can be entered into the instrument via the keyboard or directly from an office computer.① Then the surveyor guides the person holding the prism along the line of computed direction until the distance to the point to be located agrees with the computed distance. All displayed outputs can also be recorded or stored in electronic field book for further calculations in a computer. Total stations allow the measurement of many points on a surface being observed within a very short time range.

Robotic Total Station

A late 1980s adaption of the total station is the addition of servo motors to drive both the horizontal and the vertical motions of these instruments. For all the complex electronics inside a robotic total station, the motion is still provided by simple servo motors with a reduction gear system. The end result must be lightweight, durable and fast and have sub-second positioning accuracy.

When those total stations have been designed with automatic target recognition (ATR) function, they allow the user to automatically track, measure and record targets. Current technology provides robotic (motorized) total stations that are able to measure angles with an accuracy of ±0.5″ and distances with an accuracy of ±1 mm +1 ppm to a range of 3,500m. Latest models are capable of searching automatically for targets and then locking onto them precisely, turning angles automatically to designated points using the uploaded coordinates of those points, repeating angles by automatically double-centering, and even equipped with automatic data transfer systems.② These instruments, when combined with a remote controller held by the prism surveyor, enable the survey to proceed with a reduced need for personnel. All these characteristics make the robotic total stations very useful for geomatics engineering tasks.

Using a robotic total station with ATR, first, the telescope must be pointed roughly at the target prism—either manually or under software control—and then the instrument does the rest. The ATR module is a digital camera that notes the offset of the reflected laser beam, permitting the instrument then to move automatically until the cross hairs have electronically set on the point precisely. After the point has been precisely "sighted", the instrument can then read and record the angle and distance. Reports indicate that the time required this process is only one-third to one-half of the time required to obtain the same results by conventional total station techniques.

With the proper ATR-based instrument Leica TCA2003, the surveyor will be able to handle new applications and address existing jobs with a different spin.③ Because of the co-axial target detection system and the use of conventional EDM prisms, daily operation will remain unchanged. What will change is the speed with which data is collected. Topographical surveys are automated by putting the TCA in Auto-record mode—the instrument follows the rod person and automatically records a point at specified distances, time intervals, or whenever the rod is held steady for more than a certain time. Take advantage of the measuring speed and have multiple rod people on larger jobs. The TCA2003 can even turn sets of angles while the user prepares for the next traverse point.

Take automation a step further and do some "no-man" surveying. Robotic total stations are already being used in hazardous areas to provide continuous monitoring of structures or processes. Certain sites link measurement systems with civil defense agencies and law enforcement groups. An offspring of "no-man" surveying is machine guidance. TCAs guide road headers, tunneling moles and paving machinery. <u>A surveyor is in charge of installing the robotic survey station at predetermined locations and lets the robotic unit inform machine operators where they are relative to design information.</u>④ The machine operator reads the machine position from a small display receiving position updates from a base station. The base station can be robotic, GPS or a combination of a number of sensors.

Words and Expressions

prism [ˈprɪzəm] n. [物]棱镜
depress [dɪˈpres] vt. 压下；压低；使沮丧
layout [ˈleɪaʊt] n. 放样；规划；设计；(工厂等的)布局图
servo [ˈsɜːvəʊ] n. 伺服；伺服系统
gear [ɡɪə(r)] n. 齿轮；传动装置
durable [ˈdjʊərəbl] adj. 持久的；耐用的
motorize [ˈməʊtəraɪz] vt. 使机动化；使摩托化
module [ˈmɒdjuːl] n. 模数；模块；登月舱；指令舱
hazardous [ˈhæzədəs] adj. 危险的；冒险的；碰运气的
mole [məʊl] n. 全断面掘进机
paving [ˈpeɪvɪŋ] n. 衬砌；铺路
tunneling mole 隧道掘进机
predetermine [ˌpriːdɪˈtɜːmɪn] v. 预定；预先确定

☞ Notes：

① 在施工放样测量中，确定控制点到施工点方向和距离所需的数据可以通过键盘输入到仪器中或直接从办公室电脑传到仪器中。
② 最新的样机可以自动搜索目标，精确锁定目标，(可以)根据加载的设计点的坐标，自动旋转角度到点所在的方向，(可以)自动二次对中重复转换角度，甚至(还给这些样机)安装了数据自动转换系统。
③ 利用基于目标自动识别的徕卡仪器TCA2003，测量员能够应对新的应用需求，以更先进的方式从事当前的工作。
④ 测量员负责将测量机器人安装在预定的位置上，由测量机器人告知仪器操作员仪器和

设计位置间的相互关系。

Terms Highlights

instrument of geomatics engineering 测绘仪器
geodetic instrument 大地测量仪器
optical theodolite 光学经纬仪
electronic theodolite 电子经纬仪
total station 全站仪
robotic (motorized) total station 智能型全站仪；测量机器人
geo-robot 测量机器人
gyroscopic theodolite 陀螺经纬仪
data recorder 电子手簿；数据采集器
data transfer 数据转换
automatic target recognition (ATR) 目标自动识别
circular encoders 编码度盘
servo motors 伺服马达；伺服继动器
remote controller 远距离控制器
optical plummet 光学对中器

Unit 8　Errors in Measurement

Measurements are defined as observations made to determine unknown quantities. They may be classified as either direct or indirect. A direct measurement is one where the reading observed represents the quantity measured, without a need to add, take averages or use geometric formulas to compute the value desired. Determining the distance between two points by making a direct measurement using a graduated tape is an example of direct measurement. An indirect measurement requires calculation and can be determined from its mathematical relationship to direct measurements when it is not possible or practical to make direct measurements.① For example, station coordinates can be mathematically computed by measuring angles and lengths of lines between points directly. Therefore the indirect measurements (computed station coordinates) contain errors that were present in the original direct observations and propagated (distributed) by the computational process. This distribution of errors is known as error propagation. Also, it is the indirect nature of measurements that forces the need to often apply some rather sophisticated mathematical procedures to analysis of errors and thus determine a "best value" to represent the size of the quantity.

It can be stated unconditionally that all measurements, no matter how carefully executed, will contain error, and so the true value of a measurement is never known, and the exact sizes of the errors present are always unknown. Even with the most sophisticated equipment, a measurement is only an estimate of the true size of a quantity. This is because the instruments, as well as the people using them are imperfect, because the environment in which the instruments and people operate influences the process, and because the behavior of people, instruments, and the environment cannot be fully predicted. However, measurements can approach their true values more closely as better equipment is developed, environmental conditions improve and observer ability increases, but they can never be exact.

By definition, an *error* is the difference between a measured value for any quantity and its true value. The sources of errors fall into three broad categories which are described as follows:

Instrumental Errors. These errors are caused by imperfections in instrument construction or adjustment. For example, the divisions on a theodolite or total station instrument may not be spaced uniformly. These error sources are present whether the equipment is read manually or digitally.

Natural Errors. These errors are caused by variation in the surrounding environment conditions, such as atmospheric pressure, temperatures, wind, gravitational fields, and magnetic fields, etc.

Personal Errors. These errors arise due to limitations in human senses, such as the ability to read a micrometer or to center a level bubble. The sizes of these errors are affected by personal ability to see and by manual dexterity. These factors may be influenced further by temperature, insects, and other physical conditions that cause humans to behave in a less precise manner than they would under ideal conditions.

From the discussion thus far it can be stated with absolute certainty that all measured values contain errors, whether due to lack of refinement in readings, instabilities in environmental conditions, instrumental imperfection or human limitations. Some of these errors result from physical conditions that cause them to occur in a systematic way, whereas others occur with apparent randomness. Accordingly, errors are classified as either systematic or random. But before defining systematic and random errors, it is helpful to define mistakes. These three terms are defined as follows:

1. Mistakes. Mistakes or blunders (gross errors) actually are not errors because they usually are so gross in magnitude compared to the other two types of errors. Carelessness, inattention, improper training, bad habits, poor judgment, adverse measuring or observing conditions, and various negative attitudes and emotions are the traces or the common reasons for mistakes.[2] They are not classified as errors and must be removed from any set of observations. Typical examples of mistakes are omitting a whole tape length when measuring distance, sighting the wrong target in a round of angles, writing down 27.55 for 25.75 in recording. Therefore great care must be taken to obviate them.

Mistakes will never be completely eliminated from measurements, but surveyor's careful, attentive, conscientious attitude can reduce the mistakes in most cases. Through proper training and development of good work habits, development and maintenance of positive attitudes, and understanding the theory and practice involved with the variable being measured, mistakes can be controlled and practically eliminated.

2. Systematic Errors. Systematic errors are defined as those errors whose magnitude and algebraic sign can be calculated and applied as a correction to the measured quantity, or these errors follow some physical law and thus can be predicted. Some systematic errors are removed by some correct measurement procedures (e.g., balancing backsight and foresight distances in differential leveling to compensate for earth curvature and refraction). Others are removed by deriving corrections based on the physical conditions that were responsible for their creation (e.g., applying a computed correction for earth curvature and refraction on a trigonometric leveling observation).[3]

Surveyors should know how to deal with systematic errors. The first requirement is to recognize and accept the possible existence of errors. Next, identify the various sources that might be affecting a reading systematically, then, determine what the "system" is. Is it a constant, linear, or in proportion to the size of the quantity being measured? Or, does it follow some other mathematical relationship? Is there some physics involved? Once systematic errors discovered and quantified, the errors can be essentially compensated by certain processes of measuring or corrected to reduce their effect. Careful calibration of all instruments is an essential part of controlling systematic errors.

3. Random Errors. Random (also known as accident) errors are introduced into each measurement mainly because of human and instrument imperfections as well as uncertainties in determining the effects of the environment on measurements. After all mistakes and systematic errors have been removed from the measured values, the random errors remain. In general, random errors are unavoidable and relatively small. They usually do not follow any physical law, but follow random patterns, or the laws of "chance". They have unknown signs and are as likely to be negative or positive. The magnitude of such an error is unknown, but it can be dealt with and estimated according to the mathematical laws of probability. Examples of random errors are (a) imperfect centering over a ground point during distance measurement with an EDM instrument, (b) bubble not centered at the instant a level rod is read, and (c) small errors in reading graduated scales.

Understanding the nature of random errors helps to understand why random errors are never really fully corrected, since the observation of the physical phenomena contains personal, random errors. Thus, measurements have "uncertainties" or random errors that remain unquantifiable. Random errors are dealt with by controlling or managing them. It is a quality control process. They cannot be corrected or eliminated, only minimized and controlled.

Words and Expressions

geometric [ˌdʒiːəˈmetrɪk] *adj.* 几何的；几何学的
graduated [ˈɡrædʒueɪtɪd] *adj.* 分度的；分级的
propagate [ˈprɒpəɡeɪt] *v.* 传播；宣传
error propagation 误差传播
atmospheric [ˌætməsˈferɪk] *adj.* 大气的
gravitation [ɡrəˈvɪteɪʃn] *n.* 地心吸力；引力作用
micrometer [maɪˈkrəʊmɪtə(r)] *n.* 测微计；千分尺
dexterity [dekˈsterəti] *n.* 灵巧；机敏
blunder [ˈblʌndər] *n.* 粗差；大错；失误
magnitude [ˈmæɡnɪtjuːd] *n.* 数量；巨大；广大；量级
adverse [ædˈvɜːs] *adj.* 不利的；敌对的；相反的
obviate [ˈɒbvieɪt] *vt.* 消除；排除（危险、障碍等）；回避；预防；避免
eliminate [ɪˈlɪmɪneɪt] *vt.* 消除；排除
conscientious [ˌkɒnʃɪˈenʃəs] *adj.* 尽责的

maintenance [ˈmeɪntənəns] *n.*维护；保持；生活费用；赡养
quantify [ˈkwɒntɪfaɪ] *vt.*量化；确定数量
calibration [kælɪˈbreɪʃn] *n.*校准；标度；刻度

☞ Notes：

① 当不能进行直接测量或实际测量时，间接测量需要通过建立与直接测量值的数学关系计算得到。
② 粗心、怠慢、训练不当、习惯不好、判断错误、不利的观测条件以及多种消极态度及情绪通常是粗差产生的原因。
③ 其他的系统误差则可以基于误差产生的物理条件施加改正数来消除（例如在三角高程测量中施加一个计算的地球曲率和大气折射的改正数）。

Terms Highlights

law of probability 概率论
theory of error 误差理论
true error 真误差
observation error 观测误差
instrumental error 仪器误差
personal error 人为误差
gross error 粗差
systematic error 系统误差
random error, accident error 随机（偶然）误差
probable error 或然误差
constant error 常差
average error 平均误差
absolute error 绝对误差
relative error 相对误差
error of closure, closing error, closure 闭合差
error propagation, propagation of error 误差传播
error test 误差检验
gross error detection 粗差检验
tolerance 限差
limit error 极限误差
error of focusing 调焦误差
horizontal refraction error 水平折光差
index error of vertical circle 竖盘指标差

Unit 9　Basic Statistical Analysis of Random Errors

Random errors are those variables that remain after mistakes are detected and eliminated and all systematic errors have been removed or corrected from the measured values. They are beyond the control of the observer. So the random errors are errors the occurrence of which does not follow a deterministic pattern. In mathematical statistics, they are considered as stochastic variables, and despite their irregular behavior, the study of random errors in any well-conducted measuring process or experiment has indicated that random errors follow the following empirical rules:

(1) A random error will not exceed a certain amount.

(2) Positive and negative random errors may occur at the same frequency.

(3) Errors that are small in magnitude are more likely to occur than those that are larger in magnitude.

(4) The mean of random errors tends to zero as the sample size tends to infinite.

In mathematical statistics, random errors follow statistical behavioral laws such as the laws of probability. A characteristic theoretical pattern of error distribution occurs upon analysis of a large number of repeated measurements of a quantity, which conform to normal or Gaussian distribution.[①] The plot of error sizes versus probabilities would approach a smooth curve of the characteristic bell-shape. This curve is known as the normal error distribution curve. It is also called the probability density function of a normal random variable. It is important to notice that the total area of the vertical bars for each plot equals 1. This is true no matter the value of n (the number of single combined measurements), and thus the area under the smooth normal error distribution curve is equal to 1. If an event has a probability of 1, it is certain to occur, and therefore the area under the curve represents the sum of all the probabilities of the occurrence of errors.

A number of properties that relate a random variable and its probability density function are useful in our understanding of its behavior. Mean and standard deviation are two most popular statistical properties of a random variable. Generally, a random variable which is normally distributed with a mean and standard deviation can be written in symbol form as $N(\mu,\sigma^2)$.[②] They can be explained as follows:

Mean: The most commonly used measure of central tendency is the mean of a set of data (a sample). The concept of mean refers to the most probable value of the random variable. It is also called by any of the several terms—expectation, expected value, mean or average. The mean is defined as

$$\bar{x} = \frac{1}{n} \sum_{i=1}^{n} x_i$$

where x_i are the observations, n is the sample size, or total number of observations in the sample, and $\bar{x}$ is the mean which is also called most probable value (MPV). The MPV is the closest approximation to the true value that can be easily achieved from a set of data. It can be shown that the arithmetic mean of a set of independent observations is an unbiased estimate of the mean μ of the population.

Standard deviation is a numerical value indicating the amount of variation about a central value. In order to appreciate the concept upon which indices of precision devolve one must consider a measure that takes into account all the values in a set of data. Such a measure is the deviation from the mean $\bar{x}$ of each observed value x_i i. e. $(x_i - \bar{x})$, and the mean of the squares of the deviations may be used, and this is called the *variance* σ^2,

$$\sigma^2 = \frac{1}{n} \sum_{i=1}^{n} (x_i - \mu)^2$$

where μ is the mean (expectation) of the population. The square root of the variance is called *standard deviation* σ. Theoretically the standard deviation, which is the value on the X axis of the probability curve that occurs at the points of inflecxion of the curve, is obtained from an infinite number of variables known as the population.[3] In practice, however, only a sample of variables is available and S is used as an unbiased estimator. Account is taken of the small number of variables in the sample by using $(n-1)$ as the divisor, which is referred to in statistics as the *Bessel correction*; hence, variance is

$$S^2 = \frac{1}{n-1} \sum_{i=1}^{n} (x_i - \bar{x})^2$$

To obtain an index of precision in the same units as the original data, therefore the square root of the variance is used, and this is also called the *standard deviation* S. The standard deviation is the measure of the dispersion or spread of the random variable.

A survey measurement, such as a distance or angle, after mistakes are eliminated and systematic errors corrected, is a random variable. If a distance is measured 20 times, it is not unusual to get values for each of the measurements that differ slightly from its true value that is never known. So owing to random variability, an error was defined as the difference between a random variable, the measured value (observation) and the constant, the true value i. e., error = measured value − true value. And a correction (residual), which is the negative of the error in practice, was defined as correction between the MPV and measured value i. e., correction = MPV − measured value.

When the so-called true values are available to compare with calculated values, the mean square error (MSE) is given by

$$\text{MSE} = \frac{1}{n} \sum_{i=1}^{n} (x_i - \tilde{x})^2$$

in which x_i is the measured value, $\tilde{x}$ is the true value and n is the number of measurements.

Propagation of errors (or error propagation): Much data in surveying is obtained indirectly from various combinations of observations. For instance the coordinates of a line are a function of its length and bearing. As each measurement contains an error, it is necessary to consider the combined effect of these errors on the derived quantity. <u>Error propagation is one of the many aspects of analyzing errors. It is the mathematical process used to estimate the expected random error in a computed or indirectly measured quantity, caused by one or more identified and estimated random errors in one or more identified variables that influence the precision of the quantity.</u>④

The general procedure is to differentiate with respect to each of the observed quantities in turn and sum them to obtain their total effect. Thus if $Z=f(x_1,x_2,\cdots,x_n)$, and each independent variable changes by a small amount (an error) $\Delta x_1, \Delta x_2, \cdots, \Delta x_n$, then Z will change by a small amount equal to ΔZ, obtained from the following expression:

$$\Delta Z = \frac{\partial f}{\partial x_1}\Delta x_1 + \frac{\partial f}{\partial x_2}\Delta x_2 + \cdots + \frac{\partial f}{\partial x_n}\Delta x_n$$

in which $\frac{\partial f}{\partial x_i}$ is the partial derivative of f with respect to x, etc. Δ is used to replace the partial symbol d. As the observations are considered independent and uncorrelated, the variance σ_Z^2 is therefore

$$\sigma_Z^2 = \left(\frac{\partial f}{\partial x_1}\right)^2 \sigma_{x_1}^2 + \left(\frac{\partial f}{\partial x_2}\right)^2 \sigma_{x_2}^2 + \cdots + \left(\frac{\partial f}{\partial x_n}\right)^2 \sigma_{x_n}^2$$

which is the general equation for the variance of any function which is called the rationale of Error Propagation. This equation is very important and is used extensively in surveying for error analysis.

Words and Expressions

statistical [stəˈtɪstɪkl] adj.统计的；统计学的
mathematical statistics 数理统计
stochastic [stɒˈkæstɪk] adj.随机的
stochastic variable, random variable 随机变量
irregular [ɪˈregjʊlə(r)] adj.不规则的；无规律的
empirical [ɪmˈpɪrɪkl] adj.完全根据经验的；经验主义的；先验的
mean [miːn] n.平均数；中间；中庸
probability [ˌprɒbəˈbɪlətɪ] n.概率；可能性；或然性
probability curve 概率曲线；或然率曲线；几率曲线
distribution [ˌdɪstrɪˈbjuːʃn] n.分配；分发；配给物；区分；分类
normal [ˈnɔːm(ə)l] n.正态；正规；常态；[数]法线

versus [ˈvɜːrsəs] *prep.* 与……相对；对抗（指诉讼、比赛中）
approximation [əˌprɒksɪˈmeɪʃn] *n.* [数]近似值；接近；靠近
arithmetic [əˈrɪθmətɪk] *n.* 算术；算法
unbiased [ʌnˈbaɪəst] *adj.* 无偏的；没有偏见的
indices [ˈɪndɪsiːz] *n.* index的复数；[数]指数；指标；（刻度盘上的）指针
inflecxion [ɪnˈflekʃ(ə)n] *n.* [数]拐点
divisor [dɪˈvaɪzər] *n.* 除数；约数
Bessel correction 贝塞尔改正
dispersion [dɪˈspɜːʒn] *n.* [数]离差；差量；散布；驱散；传播；散射
residual [rɪˈzɪdjʊəl] *adj.* 剩余的；残留的
differentiate [ˌdɪfəˈrenʃɪeɪt] *v.* 区别；区分
derivative [dɪˈrɪvətɪv] *n.* 导数；微商
rationale [ˈræʃnəl] *n.* 基本原理

☞ **Notes：**

① 对某个量进行大量的重复测量和分析，可以得到误差分布特有的理论模式，它遵循正态分布或高斯分布。
② 一般来说，一个服从正态分布、具有期望值和标准差的随机变量，可以用符号表示为 $N(\mu, \sigma^2)$ 形式。
③ 从理论上讲，标准差是概率分布曲线拐点处对应的 X 轴上的值，由无穷多个变量（称为总体）获得。
④ 误差传播是误差分析中的一种，是用于估计计算量或间接观测量随机误差的一种数学处理方法。这些待求量的误差是由影响其精度的单个或多个变量所带的确定的或估计得到的随机误差所引起的。

Terms Highlights

error distribution 误差分布
normal distribution 正态分布
Gaussian distribution 高斯分布
probability density function 概率密度函数
normal random variable 正态随机变量
normal error distribution curve 正态误差分布曲线
most probable value (MPV) 最或然值
expectation, expected value 期望值
unbiased estimate 无偏估计
index of precision 精度指标
variance 方差
standard deviation 标准差

mean square error(MSE) 中误差
mean square error of angle observation 测角中误差
mean square error of side length 边长中误差
mean square error of a point 点位中误差
mean square error of azimuth 方位角中误差
mean square error of coordinate 坐标中误差
mean square error of height 高程中误差
variance of unit weight 单位权方差，方差因子
error ellipse 误差椭圆

Unit 10　Accuracy and Precision

Can you make a measurement that's very precise, but not very accurate? Can a number be accurate, but not very precise? Let's find out the difference between these two terms; you'll see that precision and accuracy are really two different things. A measurement can be precise but inaccurate, as well as accurate but imprecise. For example, if a measurement was made with much care using a highly refined instrument, repeated readings of the same quantity would agree closely and thus precision would exist. But if the instrument contained one or more undetected, uncorrected systematic errors, the results would be inaccurate. In contrast, it is possible that the mean of several repeated measurements of this same quantity, using a less refined (but calibrated) method, could be closer to the true value and thus this procedure would yield more accurate results even though there was less agreement among the readings.

Perhaps the easiest way to illustrate the difference between accuracy and precision is to use the analogy of a marksman, to whom the "truth" represents the bull's-eye.①

Precision: The degree of refinement in the performance of an operation, or the degree of perfection in the instruments and methods used to obtain a result, or an indication of the uniformity or reproducibility of a result.② Precision relates to the quality of an operation by which a result is obtained, and is distinguished from accuracy, which relates to the quality of the result. Figure 1 shows uniformity achieved by the marksman who is skilled by very small scatter. It illustrates the excellent precision. However, as the shots are far from the center, caused by the bent sight (systematic error), they are completely inaccurate. With the knowledge gained by observation of the results, the marksman can apply a systematic adjustment (aim higher and to the left of his intended target, or have his equipment adjusted) to achieve more accurate results in addition to the precision that his methodology and equipment have already attained.③ Such a situation can arise in practice when a piece of EDM equipment produces a set of measurements all agreeing to a few millimeters (high precision) but, due to an operation fault and lack of calibration, the measurements are all incorrect by several meters (low accuracy).

Accuracy: The degree of conformity of a final measured value, with respect to the true value as defined by accepted standard (the "truth"). Figure 2 shows the shots achieved by the marksman with a wide scatter. It illustrates that the bent sight is now corrected, i.e., the systematic errors are minimized, and the marksman has approached the "truth", although without great precision. It may be that the marksman will need to

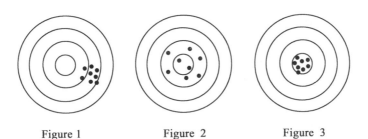

Figure 1   Figure 2   Figure 3

change the equipment or methodology used to obtain the result if a greater degree of precision is required, as he has reached the limitations associated with his equipment and methodology.

In Figure 3, the shots are clustered near the center of the target. It represents results indicating both accuracy and precision. In this case, the marksman has probably made one of the systematic adjustments that were indicated by his attainment of precision without accuracy. The degree of precision has not changed greatly, but its conformity with the "truth" has improved over the results obtained in Figure 1. The scatter is, of course, due to the unavoidable random errors.

So from the analysis of precision and accuracy, several important facts should be mentioned as follows. (1) Scatter is an "indicator of precision"; the wider the scatter of a set of results about the mean, the less reliable they will be compared with results having a small scatter. (2) Precision must not be confused with accuracy; the former is a relative grouping without regard to the nearness to the truth, whilst the latter denotes absolute nearness to the truth.[④] (3) Precision may be regarded as an index of accuracy only when all sources of error, other than random errors, have been eliminated.

An example that explains the difference between precision and accuracy better than any other in surveying has to do with error of closure in traversing. Many surveyors seem to think that error of closure checks the accuracy of the work. Wrong! Error of closure primarily checks the precision, not the accuracy. It checks accuracy only in that it can find blunders. But, since it cannot detect systematic errors in the distances, it cannot fully check accuracy.

In surveying, the need for greater precision usually leads to greater costs. To obtain a higher degree of precision, it may be necessary to use more sophisticated (costly) equipment or a more time-consuming methodology. The surveyor must determine what methodology and resultant precision is needed to achieve the accuracy required for a task at hand.

Accuracy is telling the truth … Precision is telling the same story over and over again.

Words and Expressions

refined [rɪˈfaɪnd] *adj.* 精制的；优雅的；精确的
calibrate [ˈkælɪbreɪt] *v.* 校准
marksman [ˈmɑːksmən] *n.* 射手；神射手
bull's-eye [ˈbulzai] *n.* 靶心
uniformity [ˌjuːnɪˈfɔːmɪtɪ] *n.* 一致；一式；均匀
reproducibility [ˌriːprəˌdjuːsɪˈbɪləti] *n.* 重复能力；再现性
scatter [ˈskætər] *v.* 分散；散开；撒开；驱散
bent [bent] *n.* 倾向；歪
methodology [mɪˈθɒdɒlɒdʒɪ] *n.* 方法学；方法论
conform [kənˈfɔːm] *vt.* 使一致；使遵守；使顺从
conformity [kənˈfɔːmɪtɪ] *n.* 一致；符合
cluster [ˈklʌstə(r)] *vi.* 丛生；成群
indicator [ˈɪndɪkeɪtər] *n.* 指示器；[化] 指示剂
confuse [kənˈfjuːz] *vt.* 使糊涂；搞乱
consume [kənˈsjuːm] *vt.* 消耗；消费；消灭；吸引

☞ Notes：

① 或许阐明准确度和精度两者区别最简单的方法是用射手（射击）来进行类比，这里"真值"代表靶心。
② 精度：精度是完成一次观测的精确程度，或者说是获得一个观测结果所用的仪器和方法的完善程度，或者是每个观测结果的一致性和重现能力的指标。
③ 对结果进行观察可以知道，基于现有方法和设备取得的精度，射手可以施加系统校正（瞄准目标比原来的高且偏左一点，或者校准射击装备）来获得更准确的结果。
④ 精度不可与准确度混淆；前者是指一个相对聚集的程度，不考虑其与真值的接近程度；而后者表示的是与真值的绝对接近程度。

Unit 11 Least-Squares Adjustment

Whenever the surveyor conducts a field survey, no matter how simple or complex, he invariably makes more measurements than are absolutely necessary to locate the points in the survey. A line taped in two directions introduces one measurement more that is necessary to establish the length of the line. Measuring all three angles of a triangle introduces one superfluous measurement. These extra measurements are termed redundant measurements. <u>Least-squares adjustment is a mathematical and statistical technique for dealing with the optimal combination of redundant measurements together with the estimation of unknown parameters.</u>[①] The least-squares adjustment is rigorously based on the theory of mathematical probability, whereas in general, the other methods do not have this rigorous base. In a least-squares adjustment, the following condition of mathematical probability is enforced: The sum of the square of the errors times their respective weights are minimized. In surveying, errors in measurements conform to the laws of probability, and they follow the normal distribution theory. Thus they should be adjusted in a manner that follows these mathematical laws. A mathematical model for adjustment is composed of two parts: a functional model and a stochastic model.

Mathematical Model

A *functional model* describes the geometric or physical characteristics of the survey problem. In adjustment computations a functional model is an equation that represents or defines an adjustment condition. It must either be known, or assumed. If the functional model represents the physical situation adequately, the observation errors can be expected to conform to the normal distribution curves. For example, suppose that we are interested in the shape of a plane triangle. All that is required for this operation is to measure two of its angles, and the shape of the triangle will be uniquely determined. However, if we were to decide, for safety's sake, to measure all three angles, any attempt to construct such a triangle will immediately show inconsistencies among the three observed angles. In this case the model simply is that the sum of the three angles must equal 180°. If three observations are used in this model, it is highly unlikely that the sum will equal exactly 180°. Therefore, when redundant observations, or more observations than are absolutely necessary, are acquired, these observations will rarely fit the model exactly. Intuitively, this results from something characteristic to the observations and makes them inconsistent in the case of redundancy. Of course, we first need to be sure of the adequacy of the model (it is a plane triangle and not spherical or spheroidal, for example). Then, we need to express the quality of the measurements before we seek to adjust the observations to fit the model. So from above, a well-known mathematical model states that the sum of angles in a plane triangle is 180°. This model is adequate if the survey is limited to a small region such as the plane survey.

The determination of variances, and subsequently the weights of the observations, is known as the *stochastic model* in a least-squares adjustment which describes the statistical properties of all the elements or represents a way to enter information about the precision of the observations involved in the functional model.② The importance of the stochastic model is often overlooked and undervalued. As a general rule, if the stochastic model contains misleading information, the adjustment and conclusions drawn from the adjustment can be unreliable. The stochastic model is represented by the variance-covariance matrix (weighting matrix) of the observations. It is crucial to the adjustment to select a proper stochastic (weighting) model since the weight of an observation controls the amount of correction it receives during the adjustment. However, development of the stochastic model is important not only to the weighted adjustment. When doing an unweighted adjustment, all observations are assumed to be of equal weight, and thus the stochastic model is created implicitly.

Adjustment Methods

There are two adjustment methods: the conditional and parametric adjustments. In the conditional adjustment, geometric conditions are enforced, upon the observations and their residuals. So the conditional adjustment is called direct adjustment. Examples of conditional adjustment are: (1) the sum of angles in a polygon is $(n-2) \cdot 180°$, where n is the number of angles in the polygon; (2) the sum of the angles in the horizon at any station equals 360°; (3) in a closed traverse, the algebraic sum of the departures should equal the difference between the X coordinates at the beginning and the ending stations of the traverse, similarly, the algebraic sum of the latitudes should equal the difference between the Y coordinates at the beginning and the ending stations of the traverse.③

When performing a parametric adjustment, observations are expressed in items of unknown parameters that were never measured directly. So the parametric adjustment is sometimes called indirection adjustment, in which the corrections are stated as functions of indirectly determined values of parameters of the measurements. For example, the well-known coordinate equations are used to model the measured angles, direction and distances in a traverse. The adjustment yields the most probable values for the coordinates (parameters), which in turn enable the most probable values for the adjusted observations to be computed.

A primary objective in an adjustment is to ensure that all observation used to find the most probable values for the unknowns in the model. In the least-squares adjustment, no matter conditional or parametric, the geometric checks at the end of the adjustment are satisfied and the same adjusted observations are obtained. In complicated networks, it is often difficult and time consuming to write the equations to express all the conditions that must be met for a conditional adjustment. Therefore, parametric adjustment is becoming very popular, which generally leads to larger systems of equations but is straightforward in its development and solution and, as a result, is well suited to computers.④

Words and Expressions

adjustment [əˈdʒʌstmənt] n. 平差；调整；调节
invariably [ɪnˈveərɪəbli] adv. 不变地；总是
superfluous [sjuːˈpɜːfluəs] adj. 多余的；过剩的；过量的
optimal [ˈɒptɪməl] adj. 最佳的；最理想的
parameter [pəˈræmɪtər] n. 参数；参量
rigorously [ˈrɪɡərəsli] adv. 严格地；严密地；严厉地
enforce [ɪnˈfɔːrs] vt. 强迫；执行；坚持；加强
inconsistency [ˌɪnkənˈsɪst(ə)nsi] n. 矛盾
unlikely [ʌnˈlaɪkli] adj. 未必的；不太可能的；靠不住的
rarely [ˈreəli] adv. 很少地；罕有地
intuitively [ɪnˈtjuːɪtɪvli] adv. 直觉地；直观地
overlook [ˌəʊvərˈlʊk] vt. 没注意到；俯瞰；耸出；远眺
undervalue [ˌʌndərˈvæljuː] v. 低估
mislead [ˌmɪsˈliːd] vt. 误导
implicitly [ɪmˈplɪsɪtli] adv. 含蓄地；暗中地
polygon [ˈpɒlɪɡɒn] n. [数] 多角形；多边形
horizon [həˈraɪzn] n. 地平线；范围；视野
straightforward [ˌstreɪtˈfɔːwəd] adj. 直截了当的；直观的；正直的；坦率的

☞ Notes：

① 最小二乘平差是一种处理多余观测值及确定未知参数最优解的数理统计方法。
② 在最小二乘平差中，确定方差和观测值的权称为随机模型。这种随机模型描述了所有元素(观测值)的统计特性，或者说它描述了函数模型中的观测值所具有的精度信息。
③ 在附合导线中，(观测得到的)横向坐标差之和应等于导线起始点和终点在 X 方向上的坐标差，同理，观测的纵向坐标差之和也等于导线起点和终点在 Y 方向上的坐标差。
④ 因此，参数平差变得很受青睐，总体而言，尽管这种平差方法会产生更多的方程，但

编程和解算都很直观，因而很适合计算机计算。

Terms Highlights

least squares method 最小二乘法

least-squares adjustment 最小二乘平差

adjustment of observations, survey adjustment 测量平差

functional model 函数模型

stochastic model 随机模型

redundant observation 多余观测

weight matrix 权矩阵

normal equation 法方程

variance-covariance matrix 方差-协方差矩阵

variance-covariance propagation law 方差-协方差传播律

conditional adjustment 条件平差

direct adjustment 直接平差

conditional equation 条件方程

conditional adjustment with parameters 附参数条件平差

parametric adjustment 参数平差

indirect adjustment 间接平差

observation equation 观测方程

parametric adjustment with conditions 附条件参数平差；附条件间接平差

adjusted value 平差值

covariance function 协方差函数

inverse of weight matrix 权逆阵

adjustment of correlated observation 相关平差

adjustment of typical figures 典型图形平差

approximate adjustment 近似平差

combined adjustment 联合平差

least squares collocation 最小二乘配置法；最小二乘拟合推估法

quasi-stable adjustment 拟稳平差

rank defect adjustment 秩亏平差

rigorous adjustment 严密平差

sequential adjustment 序贯平差

weight coefficient 权系数

weight function 权函数

weight reciprocal of figure 图形权倒数

Unit 12 Geodesy Concepts

As we know, surveying is divided into two major categories: geodetic surveying and plane surveying. Geodetic surveying takes into account the true shape of the Earth whereas plane surveying treats the Earth as a flat surface. The subject of this text aims at the study of the size and shape of the Earth which refers to Geodesy. The expression "the size and shape of the Earth" has various meanings in geodesy according to the way it is used and the precision with which the Earth's size and shape is to be defined. The actual topographic surface is most apparent with its variety of landforms and water areas. This is, in fact, the surface on which actual Earth measurements are made. <u>It is not suitable, however, for exact mathematical computations because the formulas which would be required to take the irregularities into account would necessitate a prohibitive amount of computations.</u>①

The concept of geodesy should be mentioned first. From the *Concise Oxford Dictionary*: *geodesy. n. The branch of mathematics dealing with the figures and areas of the Earth or large portions of it. Encyclopedia of Science and Technology*, 2001 edition, Academic Press, 2000: Geodesy is a science, the oldest Earth (geo-) science, in fact. It was born of fear and curiosity, driven by a desire to predict natural happenings and calls for the understanding of these happenings. The classical definition, according to one of the "fathers of geodesy" reads: "Geodesy is the science of measuring and portraying the Earth's surface" [Helmert, 1880, p.3]. Nowadays, we understand the scope of geodesy to be somewhat wider. It is captured by the following definition [Vanicek and Krakiwsky, 1986, p.45]: "*Geodesy is the discipline that deals with the measurement and representation of the Earth, including its gravity field, in a three-dimensional time varying space.*"

According to the classical definition of Helmert, geodesy is the "science of the measurement and mapping of the Earth's surface". This definition has to this day retained its validity; it includes the determination of the Earth's external gravity field, as well as the surface of the ocean floor. With this definition, which as to be extended to include temporal variations of the Earth and its gravity field, geodesy may be included in the geosciences, and also in engineering sciences, e.g., National Academy of Sciences (1978).

So, we've learned that <u>Geodesy is the discipline that deals with the measurement and representation of the Earth, its gravity field and geodynamic phenomena (polar motion, Earth tides, and crustal motion) in three-dimensional time varying space.</u>②

Geodesy is primarily concerned with positioning and the gravity field and geometrical aspects of their temporal variations.

Triggered by the development of space exploration, geodesy turned in collaboration with other sciences toward the determination of the surfaces of other celestial bodies (moon, other planets). The corresponding disciplines are called selenodesy and planetary geodesy.

Geodesy may be divided into three basic subdisciplines: geometric geodesy, physical geodesy, and space geodesy.

Geometric geodesy (also called astro-geodesy) concerns with determination of the size and shape of the Earth as well as the position on the Earth's surface. For the purpose of deducing the size and shape of the Earth and the precise location of specific positions on the Earth's surface, geometric geodesy considers the geoid through the use of astrogeodetic methods.③ This aspect of the science is involved with the basic principles of the establishment of the national geodetic networks which include both horizontal and vertical controls and strictly geometrical relationships measured in various ways: astronomic positioning, triangulation, trilateration, and traverse are four traditional surveying techniques in general use for determining the exact positions of points on the Earth's surface. In recent years, modern technological developments have added several new methods utilizing artificial Earth satellites. Other methods relevant to geodetic surveying are being developed.

Physical geodesy utilizes measurements and characteristics of the Earth's gravity field as well as theories regarding this field to deduce the shape of the geoid and the Earth's size in combination with arc measurements. With sufficient information regarding the Earth's gravity field, it is possible to determine geoidal undulations, gravimetric deflections, and the Earth's flattening.④

Space geodesy (Satellite geodesy) uses satellites for geodetic purposes which were advocated and published as early as 1956. With the constant growth of space technology, the deve lopment of electronic distance measuring devices, and the perfection of electronic data processing equipment, satellites specifically equipped for geodetic purposes have been deve loped, launched, observed and the data utilized. Several observational systems, geodetic cameras, electronic ranging and Doppler Satellite Surveys were developed and improved. Some of the areas of new geodetic developments are: satellite laser ranging, lunar laser ranging, very long baseline interferometry, satellite radar altimetry, the Global Positioning System, satellite-to-satellite tracking, and inertial surveying.

The major goals of geodesy can be summarized as follows [Vanicek & Krakiwsky, 1986]:

1. Establishment and maintenance of national and global three-dimensional geodetic control networks on land, recognizing the time-variant aspects of these networks.
2. Measurement and representation of geodynamic phenomena (polar motion, Earth

tides, and crustal motion).
3. Determination of the gravity field of the Earth including temporal variations.

Words and Expressions

necessitate [nəˈsesɪteɪt] v.成为必要
prohibitive [prəˈhɪbɪtɪv] adj.禁止的；抑制的
Concise Oxford Dictionary 简明牛津词典
encyclopedia [ɪnˌsaɪkləˈpiːdiːə] n.百科全书
curiosity [kjʊriˈɒsətɪ] n.好奇心
portray [pɔːrˈtreɪ] v.描绘
validity [vəˈlɪdətɪ] n.有效性；合法性；正确性
temporal [ˈtempərəl] adj.时间的；当时的；暂时的；现世的；世俗的
National Academy of Science 国家科学院
geodynamic [dʒɪ(ː)əʊdaɪˈnæmɪk] adj.地球动力学的
phenomena [fɪˈnɒmɪnə] n.现象
trigger [ˈtrɪɡər] vt.引发；引起；触发
collaboration [kəˌlæbəˈreɪʃn] n.协作；通敌
selenodesy [selɪˈnəʊdɪzɪ] n.[天]月面测量学
planetary [ˈplænətri] adj.行星的
subdiscipline [ˈsʌbˈdɪsɪplɪn] n.(学科的)分支；分科
astro-geodesy 天文大地测量学
astro-geodetic 天文大地(测量)的
deduce [dɪˈdjuːs] vt.推论；演绎出
geoidal [ˈdʒɪːɔɪdəl] adj.大地水准面的
undulation [ˌʌnˈdjuːleɪʃn] n.起伏；波动
flattening [ˈflætɪnɪŋ] n.扁率；整平
Doppler [ˈdɒplər] adj.[亦作 d-](奥地利物理学家)多普勒的
interferometry [ɪntəfɪəˈrɒmɪtrɪ] n.干涉测量(法)
altimetry [ælˈtɪmətri] n.测高学；高度测量法(以海平面为基准)

☞ Notes:

① 然而，它不适合精确的数学计算，因为精确的计算公式必须考虑地球的不规则形状，这会使得计算工作非常复杂而无法实现。
② 由此我们可知：大地测量学是一门在三维时变空间中，测量和表达地球形状、地球重力场及各种地球动力学现象(极移、固体潮和地壳运动)的学科。
③ 为了确定地球的大小和形状以及确定地球表面上特定位置的精确坐标，几何大地测量用天文大地测量方法来确定大地水准面。
④ 有了充足的地球重力场信息，我们就能确定大地水准面高、重力偏差和地球扁率。

Terms Highlights

figure of the Earth 地球形状
gravity field 重力场
polar motion 极移
Earth tide 地球潮汐；陆潮
crustal motion 地壳运动
celestial body 天体
planetary geodesy 行星大地测量学
geometric geodesy 几何大地测量学
physical geodesy 物理大地测量学；大地重力学
satellite geodesy 卫星大地测量学
space geodesy 空间大地测量学
geodetic network 大地网
astronomic positioning 天文定位
artificial Earth satellite 人造地球卫星
geoidal undulation 大地水准面高
gravimetric deflection 重力偏差
Earth's flattening 地球扁率
satellite laser ranging(SLR) 卫星激光测距
lunar laser ranging(LLR) 激光测月
very long baseline interferometry(VLBI) 甚长基线干涉测量
satellite-to-satellite tracking 卫星跟踪卫星技术
geodetic astronomy 大地天文学

Unit 13 Geoid and Reference Ellipsoid

The Earth's physical surface is a reality upon which the surveying observations are made and points are located. However, due to its variable topographic surface and overall shape, it cannot be defined mathematically and so position cannot be computed on its surface. It is for this reason that in surveys of limited extent, the Earth is treated as flat and plane trigonometry used to define position.

If the area under consideration is of limited extent, orthogonal projection of this area onto a plane surface may result in negligible distortion.① Plane surveying techniques could be used to capture field data and plane trigonometry used to compute position. However, if the area is extended to a large area beyond limitation and treated as a flat surface, the effect of the Earth's curvature will produce unacceptable distortion. It can also be clearly seen that the use of a plane surface as a reference datum for the elevations of points is totally unacceptable.

Therefore, to represent horizontal positions and elevations on maps and charts, we need a mathematical model of the Earth which includes a set of numbers for the size and shape of the Earth. We will define a mathematical surface that approximates to the shape of the area under consideration and then fit and orientate it to the Earth's surface. Such a surface is referred in surveying as a "reference ellipsoid".

The Geoid

Since the physical surface of the Earth can't be used as a computational surface, a mean sea level surface is instinctively taken into consideration. Mean sea level (MSL) is defined as the average level of the ocean surface for all stages of the tide after long periods of observations. We use MSL as a plane upon which we can reference or describe the heights of features on, above or below the ground.

By extending the Earth's MSL through the land areas, an equipotential surface approximately at MSL would be formed. Such a surface is called the "geoid". Thus by definition, the geoid is an equipotential surface of the Earth gravity field that most closely approximates the mean sea surface.② The geoid is only a theoretical surface, which is perpendicular at every point to the direction of gravity. You can't see it, touch it or even dig down to find it. The shape of geoid can be actually measured which is based on gravity data collected worldwide. Although the gravity potential is everywhere the same and the surface is smoother than physical surface of the Earth, it still contains many irregularities which render it unsuitable mathematical location of planimetric position. These irregularities are thought to be due to mass anomalies throughout the Earth.

The geoid remains important to the surveyor, as it is the surface to which all terrestrial measurements are related. As the direction of the gravity vector (termed the vertical) is everywhere normal to the geoid, it defines the direction of the surveyor's plumb-bob line. Thus any instrument which is horizontalized by means of a spirit bubble will be referenced to the local equipotential surface. Elevations are related to the equipotential surface passing through MSL. Such elevations or heights are called orthometric heights (H) and are the linear distances measured along the gravity vector from a point to the equipotential surface as a reference datum. As such, the geoid is the equipotential surface that best fits MSL and heights in question, referred to as heights above or below MSL. It can be seen from this that orthormetric heights are datum dependent.

The Reference Ellipsoid

The ellipsoid is a mathematical surface which provides a convenient model of the size and shape of the Earth. It is represented by an ellipse rotated about its minor axis and is defined by its semi-major axis a or the flattening f. The ellipsoid is chosen to best meet the needs of a particular geodetic datum system design. Although the ellipsoid is a concept and not a physical reality, it represents a smooth surface for which formulas can be developed to compute ellipsoidal distance, azimuth and ellipsoidal coordinates.[3] Due to the variable shape of the geoid, it is not possible to have a global ellipsoid of reference for use by all countries. The best-fitting global geocentric ellipsoid is the Geodetic Reference System 1980 (GRS80), which has the following dimensions: semi-major axis is 6378137.0m and semi-minor axis is 6356752.314m.

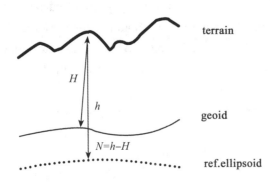

The relationship of all three surfaces which are terrain, geoid and ellipsoid is illustrated in this Figure. We note that the orthometric height H is the height with reference to the MSL, whereas the geodetic height (h) is the height of anything above the reference ellipsoid. The relation between the two kinds of heights is shown in the Figure, where the quantity N, the height of the geoid above the reference ellipsoid or the perpendicular distance between the geoid and the reference ellipsoid at a point, is usually called the geoidal height (geoid undulation). Thus, the knowledge of the geoid is necessary for transforming the geodetic to orthometric height and vice versa.④ Once we determine the geoid, we can compute the difference between the two surfaces, the ellipsoid and the geoid anywhere in the country. The expression "ellipsoidal height" for (geodetic) height of anything above the reference ellipsoid is also used comparing the acceptance of the standard geodetic term of "geodetic height".

Surveyors used to working with spirit levels have referenced orthometric heights (H) to the "average" surface of the Earth, as depicted by MSL. The surface of MSL can be approximated by the geoid. The difference between the two surfaces arises from the fact that seawater is not homogeneous and because of a variety of dynamical effects on the seawater. The height of the MSL above the geoid is called the sea surface topography (SST). It is a very difficult quantity to obtain from any measurements; consequently, it is not yet known very accurately.

GPS heights are referenced to the ellipsoid, a mathematical model that does not physically exist. This model, does not agree with mean sea level. That means the height of a point determined from GPS is not the same as its sea level elevation as determined by leveling.

The summarizing of the relationships among height systems can be illustrated below:

(1) MSL elevation is roughly equivalent to orthometric height (H), the technical name for height above the geoid. The geoid is, for all intents and purposes, the same as MSL.

(2) Geoidal height (N) is the separation between the geoid and the ellipsoid (sometimes called Geoidal separation). It can be plus or minus. A negative geoidal separation indicates that the geoid is below the ellipsoid.

(3) Ellipsoidal height (h) is the distance above or below the ellipsoid (plus or minus). Ellipsoidal height is also called geodetic height.

Words and Expressions

orthogonal [ɔːˈθɒɡənl] *adj.* 正交的；直角的；直交的
distortion [dɪˈstɔːʃn] *n.* 变形；扭转；歪扭；歪曲作用；反常；畸变
instinctively [ɪnˈstɪŋktɪvli] *adv.* 本能地
equipotential [ˌiːkwɪpəˈtenʃəl] *adj.* 等位的；有相等潜力的；等电位的
potential [pəˈtenʃl] *n.* 潜能；潜力；电压
render [ˈrendər] *vt.* 致使；呈递；归还；汇报；放弃；表演；实施
anomaly [əˈnɒməli] *n.* 不规则；异常的人或物
terrestrial [təˈrestriəl] *adj.* 陆地的
homogeneous [ˌhɒməˈdʒiːniəs] *adj.* 同类的；相似的；均一的；均匀的
dynamical [daɪˈnæmɪk(ə)l] *adj.* 动力(学)的；有力量的
intent [ɪnˈtent] *n.* 意图；目的；意向

☞ **Notes**:

① 如果所考虑的区域面积在一定范围内，则该区域正射投影到平面上所引起的变形可以忽略。
② 所以根据定义，大地水准面就是最接近于平均海水面的重力等位面。
③ 虽然这个椭球只是一个概念，并非实体，但它表示了一个平滑曲面，可以以它为基础，用公式计算椭球距离、方位角以及椭球坐标。
④ 因此，大地高转换成正高需要知道大地水准面，反之(由正高转换成大地高)亦然。

Terms Highlights

mean sea level (MSL) 平均海[水]面
geoid 大地水准面
reference ellipsoid 参考椭球
semi-major axis [of ellipsoid] 椭球长半轴；地球长半轴
semi-minor axis [of ellipsoid] 椭球短半轴；地球短半轴
flattening [of ellipsoid] 椭球扁率
eccentricity of ellipsoid 椭球偏心率
orthogonal projection 正交投影；正射投影
gravity vector 重力矢量
orthometric height(H) 正高
geodetic height, ellipsoidal height(h) 大地高
geoidal height, geoid undulation(N) 大地水准面高；大地水准面差距
sea surface topography (SST) 海面地形
quasi-geoid 似大地水准面
normal height 正常高
dynamic height 力高
height anomaly 高程异常
Geodetic Reference System 1980 (GRS80) 1980 大地测量参考系统
Krasovsky ellipsoid 克拉索夫斯基椭球

Unit 14 Datums, Coordinates and Conversions

A datum is the mathematical model of the Earth we use to calculate the coordinates and elevations on any map, chart, or survey system. Geodetic datums define the size and shape of the Earth and the origin and orientation of the coordinate systems used to map the Earth. Hundreds of different datums have been used to frame position descriptions since the first estimates of the Earth's size were made by Aristotle. But all coordinates reference some particular set of numbers for the size and shape of the Earth. For example, the Global Positioning System (GPS) is based on the World Geodetic System 1984 (WGS-84). Many countries use their own datums when they make their maps and surveys—what we call local datums.

Horizontal and Vertical Datums

In geodesy two types of datums must be considered: a "Horizontal Datum" for location, which forms the basis for the computations of horizontal control surveys in which the curvature of the Earth is considered and a "Vertical Datum" for elevation or to which elevations are referred. Vertical control networks provide elevations with reference to a surface of constant gravity force called the geoid.[①] Almost all maps and charts use mean sea level (geoid) for elevation. But they might use any of over hundred different horizontal position datums. One example of horizontal datum is a mathematical surface called a reference ellipsoid with which positional information (latitude and longitude) is referenced to. The coordinates for points in specific geodetic surveys and triangulation networks are computed from certain initial quantities (datums). Sometimes, a map will have more than one grid on it. Normally, each grid is for a different datum.

Horizontal datum and coordinates: A horizontal datum is a surface of constant values that forms the basis for the computations of horizontal control surveys. In a horizontal datum a reference ellipsoid is used as a mathematical approximation of the shape of the Earth. Five parameters are required to define a horizontal datum: two to specify the dimensions of the ellipsoid, two to specify the location of an initial point (origin), and one to specify the orientation (i.e., north) of the coordinate system. For example, the radius and flattening of the ellipsoid selected for the computations to specify the dimensions of the ellipsoid, the longitude and latitude of an initial point (origin) to specify the location and an azimuth of a line (direction) to some other (triangulation) station to specify the orientation.[2] A change in any of these quantities affects every point on the datum. For this reason, while positions within a system are directly and accurately reliable, data such as distance and azimuth derived from computations involving geodetic positions on different datums will be in error in proportion to the difference in the initial quantities.[3]

The two main horizontal datums used in the U.S. are the North American Datum of 1927 (NAD27) and the North American Datum of 1983 (NAD83). In 1986, NAD83 replaced NAD27 because the latter was found to be not accurate enough to support modern positioning activities that occur in highly accurate electronic measurement systems and satellite-based positioning systems. NAD83 is an earth-centered datum and relies on an ellipsoid (and other constants) of the Geodetic Reference System of 1980 (GRS80). It is important to note that GPS position calculations are based on the WGS84 datum (World Geodetic System of 1984), which for all practical purposes is identical to GRS80. In China, Xi'an Geodetic Coordinate System 1980 is used as a horizontal datum in which the initial point (origin) is in Shaanxi Province.

Vertical datum and heights: The zero surface, to which elevations or heights are referred, is called a vertical datum. From previous text we know the geoid is an equipotential surface of the Earth gravity field that most closely approximates the mean sea surface. At every point the geoid surface is perpendicular to the local plumb line. It is therefore a natural reference for heights — measured along the plumb line. Heights referred to the geoid are called orthometric heights, which stand in contrast to geodetic (ellipsoidal) heights, which refer to the ellipsoid. Because we cannot directly see the geoid surface, we cannot actually measure the heights above or below the geoid surface. We must infer where this surface is by making gravity measurements and by modeling it mathematically. For practical pur poses, we assume that at the coastline the geoid and the MSL surfaces are essentially the same. Nevertheless, as we move inland we measure heights relative to the zero height at the coast, which in effect means relative to MSL. Therefore we use mean sea level as a plane upon which we can reference or describe the heights of features on, above or below the ground.

Elevations are not required for most parcel mapping applications. However, since GPS is a 3D (actually 4D) measuring device, elevations are available for every point. As mentioned earlier, the GPS-derived elevation refers to the ellipsoid (ellipsoidal height), not the mean sea level (orthometric height).

Conversions

A coordinate conversion or transformation is the process of bringing a coordinate from one defined coordinate system (or zone) into another through a series of algorithms based on the latitude/longitude position of the point. Coordinate systems based on the same datum retain a perfect mathematical relationship, allowing coordinate values to be precisely transformed between them.

But the coordinates for a point on the Earth's surface in one datum will not match the coordinates from another datum for that same point. The differences occur because of the different ellipsoids used and the probability that the centers of each datum's ellipsoid is oriented differently with respect to the Earth's center. A grid shift exists between datums because each datum has a different origin.

A datum conversion is the process of bringing coordinate values referenced to one defined datum into another datum systems. Complete datum conversion is based on seven parameter transformations that include three translation parameters, three rotation parameters and a scale parameter. Simple three parameter conversion between latitude, longitude, and height in different datums can be accomplished by conversion through Earth-Centered, Earth Fixed XYZ Cartesian coordinates in one reference datum and three origin offsets that approximate differences in rotation, translation and scale.④

Words and Expressions

conversion [kənˈvɜːrʒn] n.变换；转化
origin [ˈɔːrɪdʒɪn] n.原点；起算点
orientation [ˌɔːriənˈteɪʃn] n.定位；方向；方位；倾向性；向东方
frame [freɪm] vt.构成；设计；制定
Aristotle [ˈærɪstɒtl] n.亚里士多德（古希腊大哲学家、科学家）
infer [ɪnˈfɜːr] v.推断
nevertheless [ˌnevərˈðeləs] conj.然而；不过
parcel [ˈpɑːrsl] n.小包；包裹
algorithm [ˈælɡərɪðəm] n.[数]运算法则
translation [trænsˈleɪʃn] n.[数]平移；翻译；译文；[物]转换
rotation [rəʊˈteɪʃn] n.旋转
Cartesian [kɑːˈtiːziən] adj.笛卡儿的；笛卡儿哲学的 n.笛卡儿坐标系
offset [ˈɒfset] n.偏移量；弥补；分支

☞ **Notes**：

① 高程控制网提供了以大地水准面这个重力等位面为基准参考的高程。
　with reference to：关于
② 例如，计算中用椭球的半径和扁率定义参考椭球的大小，用初始点（即原点）的经纬度来定义椭球的位置，用（原点）到某三角测量点的一条边的方位角来进行椭球定向。
③ 因此，即使我们有某参考系下准确和可靠的点位，但不同基准之间，涉及大地点位计算得到的距离和方位角的数据也会存在差异，这种差异与不同基准的参数差异成比例。
　in proportion to：与……成比率（例）
④ 不同基准之间经纬度和高程的转换可以通过简易的三参数转换实现，即通过一个参考

基准下的地心地固笛卡儿坐标和近似表示旋转、平移和缩放参数影响的三个原点平移参数来完成。

Terms Highlights

geodetic datum　大地基准
geodetic position　大地位置
geodetic coordinate system　大地坐标系统
geodetic origin　大地原点
geodetic latitude　大地纬度
geodetic longitude　大地经度
world geodetic system of 1984（WGS-84）　1984世界大地坐标系（WGS-84）
geodetic coordinate　大地坐标
rectangular grid　直角坐标网
independent coordinate system　独立坐标系
coordinate conversion　坐标变换
translation parameters　平移参数
rotation parameters　旋转参数
scale parameter　尺度参数
celestial coordinate system　天球坐标系
CIO（conventional international origin）　国际协议原点
Xi'an Geodetic Coordinate System 1980　1980西安坐标系

Unit 15 Map Projection

Map projections are attempts to portray the surface of the Earth or a portion of the Earth on a flat surface. Some distortions of conformality, distance, direction, scale, and area always result from this process. Some projections minimize distortions in some of these properties at the expense of maximizing errors in others. Some projections are attempts to only mode rately distort all of these properties. No projection can be simultaneously conformal and area-preserving.

Conformality: When the scale of a map at any point on the map is the same in any direction, the projection is conformal. Meridians (lines of longitude) and parallels (lines of latitude) intersect at right angles. Shape is preserved locally on conformal maps.

Distance: A map is equidistant when it portrays distances from the center of the projection to any other place on the map.

Direction: A map preserves direction when azimuths (angles from a point on a line to another point) are portrayed correctly in all directions.

Scale: Scale is the relationship between a distance portrayed on a map and the same distance on the Earth.

Area: When a map portrays areas over the entire map so that all mapped areas have the same proportional relationship to the areas on the Earth that they represent, the map is an equal-area map.①

Classification of Map Projection

Map projections are generally classified into four general classes according to common properties (cylindrical vs. conical, conformal vs. area-preserving, etc.), although such schemes are generally not mutually exclusive.

Cylindrical projections result from projecting a spherical surface onto a cylinder. A cylindrical projection can be imagined in its simplest form as a cylinder that has been wrapped around a globe at the equator. If the graticule of latitude and longitude are projected onto the cylinder and the cylinder unwrapped, then a grid-like pattern of straight lines of latitude and longitude would result.② The meridians of longitude would be equally spaced and the parallels of latitude would remain parallel but may not appear equally spaced anymore. In reality cylindrical map projections are not so simply constructed. The three aspects of the cylindrical projections are as follows:

- Tangent or secant to equator is termed *regular*, *or normal*. When the cylinder is tangent to the sphere contact is along a great circle (the circle formed on the surface of the Earth by a plane passing through the center of the Earth). In the secant case, the cylinder touches the sphere along two lines, both small circles (a circle formed on the surface of the Earth by a plane not passing through the center of the Earth).

- Tangent or secant to a meridian is the *transverse* aspect. When the cylinder upon which the sphere is projected is at right angles to the poles, the cylinder and resulting projection are transverse.
- Tangent or secant to another point on the globe is called *oblique*. When the cylinder is at some other, non-orthogonal, angle with respect to the poles, the cylinder and resulting projection is oblique.

Conic projections result from projecting a spherical surface onto a cone. When the cone is tangent to the sphere contact is along a small circle. In the secant case, the cone touches the sphere along two lines, one a great circle, the other a small circle. In the Conical Projection the graticule is projected onto a cone tangent, or secant, to the globe along any small circle (usually a mid-latitude parallel). In the normal aspect (which is oblique for conic projections), parallels are projected as concentric arcs of circles, and meridians are projected as straight lines radiating at uniform angular intervals from the apex of the flattened cone. Conic projections are not widely used in mapping because of their relatively small zone of reasonable accuracy. The secant case, which produces two standard parallels, is more frequently used with conics. Even then, the scale of the map rapidly becomes distorted as distance from the correctly represented standard parallel increases. Because of this problem, conic projections are best suited for maps of mid-latitude regions, especially those elongated in an east-west direction. The United States meets these qualifications and therefore is frequently mapped on conic projections.

Azimuthal (Planar) projections result from projecting a spherical surface onto a plane. When the plane is tangent to the sphere contact is at a single point on the surface of the Earth. In the secant case, the plane touches the sphere along a small circle if the plane does not pass through the center of the Earth, when it will touch along a great circle.

Miscellaneous projections include unprojected ones such as rectangular latitude and longitude grids and other examples of that do not fall into the cylindrical, conic, or azimuthal categories.

Choosing a projection is to determine: Location, Size and Shape. These three things determine where the area to be mapped falls in relation to the distortion pattern of any projection. One "traditional" rule described by Maling (Maling, 1992) says:

A country in the tropics asks for a cylindrical projection.

A country in the temperate zone asks for a conical projection.

A polar area asks for an azimuthal projection.

Implicit in these rules of thumb is the fact that these global zones map into the areas in each projection where distortion is lowest: Cylindricals are true at the equator and distortion increases toward the poles. Conics are true along some parallel somewhere between the equator and a pole and distortion increases away from this standard. Azimuthals are true only at their center point, but generally distortion is worst at the edge of the map.

For a particular map-use the map may need to be conformal, equal area, or some compromise of these. In some cases, such as navigation, conformality is absolutely necessary. In statistical mapping, equivalence is necessary. The final projection choice would seem to be a fairly straightforward function of minimized distortion and special properties.

Universal Transverse Mercator (UTM)

Mercator projection was invented in 1569 by Gerardus Mercator (Flanders) graphically. The properties of this projection are: (1) Conformal. (2) Meridians unequally spaced, distance increases away from equator directly proportional to increasing scale. (3) Loxodromes or rhumb lines are straight. (4) Used for navigation and regions near equator.

The accuracy of Transverse Mercator projections quickly decreases from the central meridian. Therefore, it is strongly recommended to restrict the longitudinal extent of the projected region to +/− 10 degrees from the central meridian.

The UTM system applies the Transverse Mercator projection to mapping the world, using 60 pre-defined standard zones to supply parameters.③ UTM zones are six degrees wide. Each zone exists in a North and South variant.

Words and Expressions

conformal [kɒnˈfɔːməl] *adj.* [数]（地图投影中）正形投影的；等角的
conformal map 保角映像；保形变换图
moderately [ˈmɒdərətli] *adv.* 适度地
equidistant [ˌiːkwɪˈdɪstənt] *adj.* 距离相等的；等距的
cylindrical [səˈlɪndrɪkl] *adj.* [计] 圆柱的
cylinder [ˈsɪlɪndər] *n.* 圆筒；圆柱体；汽缸；柱面
cylindrical projection 圆柱投影
conical [ˈkɒnɪk(ə)l] *adj.* 圆锥的；圆锥形的
cone [kəun] *n.* [数、物] 锥形物；圆锥体
conic projection 圆锥投影
mutually [ˈmʌtʃuəli] *adv.* 互相地；互助地
graticule [ˈɡrætɪkjuːl] *n.* 分成小方格以便复制的图形；格子线
secant [ˈsiːkənt] *adj.* 切的；割的；交叉的 *n.* 割线；正切

transverse [trænˈzvɜːs] *adj.* 横向的；横断的
oblique [əˈbliːk] *adj.* 倾斜的；间接的
concentric [kənˈsentrɪk] *adj.* 同中心的
apex [ˈeɪpeks] *n.* 顶点
elongate [iːˈlɒŋgeɪt] *v.* 拉长；(使)伸长；延长
qualification [ˌkwɒlɪfɪˈkeɪʃn] *n.* 资格；条件；限制；限定；赋予资格
planar [ˈpleɪnər] *adj.* 平面的；平坦的
antipodal [ænˈtɪpədəl] *adj.* 对跖的；正相反的
miscellaneous [ˌmɪsəˈleɪniəs] *adj.* 各色各样混在一起的；混杂的；多才多艺的
unprojected [ˌʌnˈprɒdʒektɪd] *adj.* 未计划的；非预料的
thumb [θʌm] *n.* 拇指
rules of thumb 经验法则
equivalence [ɪˈkwɪvələns] *n.* 同等；[化]等价；等值
Mercator [mɜːrˈkeɪtə(r)] 墨卡托(1512—1594，荷兰地理学家，地图制作家)
loxodrome [ˈlɒksədrəʊm] *n.* 斜航线；恒向线
rhumb [rʌm] *n.* 罗盘方位；罗盘方位单位
rhumb line 等角线；恒向线；无变形线；斜航

☞ **Notes**：

① 当一幅地图将所有的区域描绘在整个地图上时，所有绘图面积和他们所描述的地面面积有相同的比例关系，这种地图被称为等积地图。
② 圆柱投影可以简单地设想为一圆柱套在地球上，与地球的赤道相切。如果将经纬格网线投影到圆柱上，然后将圆柱展开就得到了直线格网状的经纬线。
③ 通用横轴墨卡托投影应用横轴墨卡托投影原理绘制世界地图，用预先定义好的60个标准投影带提供参数。

Terms Highlights

map projection 地图投影
arbitrary projection 任意投影
conformal projection 等角投影；正形投影
equidistant projection 等距投影
equivalent projection 等积投影
orthographic projection 正射投影
Mercator projection/Mercator's projection 墨卡托投影
polyconic projection 多圆锥投影
gnomonic projection 球心投影；极平投影
perspective projection 透视投影
normal projection 正轴投影
transverse projection 横轴投影
oblique projection 斜轴投影

varioscale projection 变比例投影
Universal Transverse Mercator（UTM）通用横墨卡托投影
Lambert projection/Lamber's projection 兰勃特投影
Universal Polar Stereographic projection(UPS) 通用极球面投影

Unit 16　Gravity Measurement

As known from daily experience, the most conspicuous force present on the surface of the Earth is gravity. Gravity is the force that pulls things towards the center of the Earth. Gravity affects almost everything in our lives. From clocks to hydroelectric dams, from the tides of the oceans to plant life, gravity plays an important role. Gravity governs our height and shape and keeps us from falling off the surface of the Earth.

Concepts of Gravity

Gravitation is the force that pulls or attracts all bodies in the universe towards one another. The famous law of universal gravitation, first formulated mathematically by Newton states the fact that any two physical bodies attract each other. This gravitation force, which is proportional to the product of the two masses (M and m) and inversely proportional to the square of their distance, can be expressed as follows:

$$F = G \frac{Mm}{r^2}$$

in which G is known as Newton's gravitational constant. Since the Earth is spinning, any object within or on the Earth's surface pursues a circular path as the Earth rotates on its axis. If the body is on the Earth's surface, it follows the circular path because it is held onto the Earth's surface by the gravitational force. However, the body as it pursues a circular path exerts an outward force called the centrifugal force.[①] The sum of the gravitation and the centrifugal force is called the gravity force. The gravity force is stronger on the poles than on the equator, since the centrifugal force vanishes on the pole whereas the value of the centrifugal force on the equator is obtained biggest. Even though centrifugal force is the strongest on the equator, it is about 0.35% of the gravitation force. Therefore the gravity is mainly governed by the gravitation force. The gravity causes an object to have weight and, if the object is free to move, to fall with increasing speed (i.e., the body accelerates) toward the center of the Earth. The acceleration experienced by that object as it moves toward the Earth's center is called the acceleration of gravity, which is the quantity observed when gravity measurements are made.

In study of the geometrical properties of the gravity, it is sufficient to concentrate on the acceleration. The acceleration of gravity is equal to the gravity when the mass m is one unit. The units of gravity are Gals, named after the famous physicist Galileo. One

Gal equals one centimeter per second squared (cm/s^2). The acceleration of gravity is not constant across the surface of the Earth. The differences are caused by the varying densities of sub-surface rocks, location on the surface (such as the latitude) and the elevation which is the distance from the center of the Earth. A mean magnitude of gravity on the surface of the Earth is of the order of 980.3Gal. The global range of the variations on the surface of the Earth is more than 5Gal. These variations are easily measured even with imprecise instruments. modern instruments measure accurately to within a fraction of a μ Gal (1μ Gal $= 10^{-6}$ Gal) or even higher.

From the above formula, we know the gravity is inversely proportional to the square of their distance. It is obvious that the different heights of observation points will affect the gravity. It is proved from [P. Vanicek & E. J. Krakiwsky, 1986] that the gravity magnitude decrease with the increasing height. The gravity decreases by only 1% with an increase in altitude of about 32km. Furthermore, the Earth is made from different materials of dif ferent mass with varying densities and the Earth is not spherical but oblate, the gravity field of the Earth also varies due to anomalies within the Earth and the oblateness of the Earth. All have an impact on the Earth's gravitational field.[2] Thus the different heights of observation points, the oblateness of the Earth, and the uneven lateral distribution of mass within the Earth are three sources of the gravity variations.

How to Measure Gravity

The basic way to measure gravity is to drop something and watch its trajectory as it falls. That is Newton's law. The magnitude of gravity can be measured using any of the existing kinds of gravity measurements. Two distinctly different types of gravity measurements are made: absolute gravity measurements and relative gravity measurements. If the value of acceleration of gravity can be determined at the point of measurement directly from the data observed at that point, the gravity measurement is absolute. If only the differences in the value of the acceleration of gravity are measured between two or more points, the measurements are relative.

Until the middle of the 20th century, virtually all absolute measurements of gravity were made using some type of pendulum apparatus. The most usual type of apparatus contained a number of pendulums that were swung in a vacuum. By measuring the period of the pendulums, the acceleration of gravity could be computed. In recent years, the pendulum method has been superseded by the ballistic method which is based on timing freely falling bodies. The acceleration of gravity can be determined by measuring the time taken by a body to fall over a known distance.[3]

Over the years, absolute gravity measurements have been made at only a few key sites, and these few measurements have served chiefly to establish datum and scale for relative gravity measurements. The reason for the sparsity of absolute gravity measurements is that the necessary measuring equipment is very bulky and costly, and a single measurement requires days of painstakingly careful work under near laboratory conditions. This, however, may change in the next decade or so. As absolute equipment continues to be miniaturized and made more portable, absolute gravity measurements are becoming more commonplace.

<u>Since absolute gravity measurements have been too complicated and time consuming and, until recently, could not be obtained with sufficient accuracy, relative gravity measurements have been used to establish the dense network of gravity measurements needed.</u>④ Modern relative gravity measurements are made with small, very portable, and easily used instruments known as gravimeters (gravity meters). Using gravimeters, highly accurate relative measurements can be made at a given site, known as a gravity station, in half-an-hour or less.

Since relative gravity surveys can determine only differences in gravity from point to point, every relative gravity survey must include measurements at one or more reoccupiable points where acceleration of gravity is known. Such points are called base stations. Then all gravity difference measurements are computed with respect to the known gravity value at the base station. Hence, tying a relative gravity survey to a base station establishes the "gravity datum" of that survey.

Words and Expressions

conspicuous [kənˈspɪkjuəs] adj.显著的；显而易见的
hydroelectric [ˌhaɪdrəʊɪˈlektrɪk] adj.水力电气的
dam [dæm] n.水坝；障碍
Newton [ˈnjuːt(ə)n] n.牛顿
product [ˈprɒdʌkt] n.乘积；产品；产物
spin [spɪn] v.旋转；纺；纺纱
exert [ɪɡˈzɜːt] vt.施加（压力）；努力；发挥；竭尽全力
centrifugal [ˌsentrɪˈfjuːɡl] adj.离心的
Gal [ɡæl] n.[加速度单位]伽；[重力加速度单位]加仑
Galileo [ɡəˈliːləʊ] n.伽利略（1564—1642,意大利物理学家及天文学家）
decrease [dɪˈkriːs] v.减少
oblate [ˈɒbleɪt] adj.[几何]扁平的；扁圆的
oblateness [əˈbleɪtnɪs] n.扁率；扁球形；扁圆形
trajectory [trəˈdʒektəri] n.[物]（射线的）轨道；弹道；轨迹
apparatus [ˌæpəˈrætəs] n.仪器；器械；设备
supersede [ˌsuːpərˈsiːd] vt.代替；取代；接替；紧接着……而来；[律]延期
ballistic [bəˈlɪstɪk] adj.弹道的；弹道学的
sparsity [ˈspɑːsɪti] n.稀少
bulky [ˈbʌlki] adj.大的；容量大的；体积大的
painstakingly [ˈpeɪnzteɪkɪŋli] adv.辛苦地；辛勤地；艰苦地

miniaturize [ˈmɪnətʃəraɪz] *vt.* 使小型化
portable [ˈpɔːrtəbl] *adj.* 轻便的；手提（式）的；便携式的

☞ **Notes：**

① 地球上的物体由于万有引力的作用，就会受地球吸附并沿圆形轨迹运动。然而，作圆周运动的物体会产生一个（沿径向）向外的力称为离心力。
② 此外，地球是由密度不同、质量各异的物质所组成的，其形状也不是球形而是扁球形，因而地球重力场也会随地球内部异常和地球扁率发生变化，这些都对地球引力场有影响。
③ 近年来，这种利用摆的方法被一种基于测量自由落体运动时间的弹道学方法所代替，通过测量物体下落一段已知距离所需的时间来求得重力加速度。
④ 由于绝对重力测量复杂耗时，近年来也没能获得足够满意的精度，因此采用相对重力测量来建立重力测量所需的加密网。

Terms Highlights

gravity measurement 重力测量
law of universal gravitation 万有引力定律
centrifugal force 离心力
gravitational constant 引力常数；重力常数
acceleration of gravity 重力加速度
gravitational field 重力场；引力场
absolute gravity measurement 绝对重力测量
relative gravity measurement 相对重力测量
gravimeter, gravity meter 重力仪
gravity station 重力点
gravity datum 重力基准
gravimetric baseline 重力基线
gravimetric deflection of the vertical 重力垂线偏差
gravitational potential 引力位
gravity anomaly 重力异常
gravity gradient measurement 重力梯度测量
gravity observation of Earth tide 重力固体潮观测
gravity potential 重力势，重力位
gravity reduction 重力归算

Unit 17　Optimal Design of Geomatics Network

Overview

Engineering survey control networks (geomatics networks) are established for setting out bridges, tunnels, power dams and for monitoring those artificial constructions or natural ones like slides, subsidence areas, etc.[①] Mostly the network design is done according to the experience of the surveying engineer. Taking into account the situation at the site, control points are established in areas which are regarded to be stable and from which points are observed via geodetic means. However, it may not be guaranteed that the control points remain stable. In general they have to be observed and evaluated relative to one another and to the object points in a more complex geomatics network. The design and optimization of the corresponding observation plans and configurations get more and more a topic of research with increasing importance of modern positioning methods, especially GPS observations.

The design of a geomatics network to meet a given set of specifications can involve four distinctly different types of design problem [Grafarend, 1985]:

(1) Zero-order design: The datum problem, in which both the design matrix (A) and the covariance matrix of the observables (C_L) are known, the problem is to determine the covariance matrix of the unknown position (C_{xx}). It is necessary to introduce some restrictions to the parameters to solve the datum problem. In free networks the inner datum or special strong datum are in common use. Therefore parameters and corresponding variance-covariance matrices between different datum systems may be transformed by means of s-transformations. The usually approaches which are originally valid for conversions between systems of equal defect are extended to the problems of increasing or diminishing the defect.

(2) First-order design: The select of the configuration of observations, in which the covariance matrix of the unknown position (C_{xx}) and the covariance matrix of the observables (C_L) are known. The problem is to find what geometric combination of observables (design matrix A) will yield the desired solution. The configuration problem includes the optimization of the positions of the points of a network as well as of the observation plan.

(3) Second-order design: The choice of observation weights, in which the C_{xx} and A matrix are known, what is wanted is the C_L matrix required for the observations in order to achieve the specified C_{xx} matrix. <u>The weight optimization is the problem to find optimal weights respectively or accuracies to realize the observation plan for a given configuration, which meet the requirements to the network, the problem can be solved in principle with any comfortable adjustment program by a prior simulation, where a controlled variation of weight or accuracies and the analysis of the actual adjustment result in teract.</u>[②]

(4) Third-order design: The improvements of the design in which only the C_{xx} matrix is given and both the A and C_L matrix must be solved for the problem.

Approach to the Optimal Design

There are two approaches to the solution of these design problems. They can be solved for in an iterative (trial and error) manner or by a direct (mathematical) solution. Using the iterative approach the design first solves for a C_{xx} matrix using a first approximation to a good solution to the problem. Then the difference between the computed and desired C_{xx} matrix are analyzed, the approximate solution modified, and the desired C_{xx} matrix is updated. This process continues until the solution provides a C_{xx} matrix close enough to the desired one. While this method will not yield the exact "optimum" solution (in the mathematical sense), the solution arrived at will be good enough to use in practice. <u>The principal disadvantage with this method is the time delay between the modification of the previous design and the receipt of the results of the preanalysis of the modified design. The result is a loss of time of continuity in the process that significantly hinders the efficient designing of the network.</u>[③] The iterative design work may be carried in the following manner:

(1) To begin with, a "maximum proposal" is prepared, i.e., a proposal containing all the measurements that are performable and that are realistic for inclusion from the technical and economic standpoint.

(2) This proposal is simulated and if the precision is sufficient and the reliability well attuned the measurement program is approved, otherwise it is modified.

(3) If the requirements do not satisfy, for instance if the standard errors are too large and/or there are uncontrollable measurements, the program must be expanded or the measurement accuracy must be increased. For example, it may be necessary to increase the number of rounds or to change to a more accurate theodolite. It may also be necessary to measure the distance twice. If it is rather a matter of minor local problems, the solution is to add additional measurements if it is possible to provide help for the weak parts.

(4) The program may also be overambitious. In this case, superfluous measurements

are simply excluded. It is essential not to make too many changes at the same time, but as a rule it is possible to modify in one swoop more than can be believed at first.

(5) Naturally, there may be both superfluous and uncontrollable measurements in one and the same network. Changes to remedy these problems can, however, take place simultaneously unless exactly the same part is affected in both cases.

(6) After the changes have been introduced, a new simulation is made and the whole thing is repeated until the design can be approved in its entirety.

The direct approach in which the required information is solved mathematically is intuitively more pleasing. It does, however, have several problems. The C_L matrix that results is generally fully populated and is frequently singular. Both these properties are in direct conflict with reality. Surveying observations are generally uncorrelated and the C_L matrix for them is, by definition, non-singular. When using the direct approach to solve for the design matrix other anomalies can occur, these include the absence of redundancy for some stations and the spitting of a single network into two or more independent networks, which of course does not correspond at all with reality.

Quality Control of Network

Objective criteria for network design which should be considered are those of precision, of sensitivity, of reliability and of economy. The precision can be derived from the stochastic properties of the observations applying the law of propagation of errors. In the case of adjustment, the variance covariance matrix of the parameters to be determined describes the precision behavior. The same is true for the sensitivity with regard to an expected deformation behavior. It is also based on the error distribution of the observation and on the geometry of determination of the parameters. Using terminology from quality control, the precision and the sensitivity discussed here can be considered as the quality of the design. <u>Perhaps even more important for engineering networks problems such as the monitoring problems is the quality of conformances, i. e., the fidelity with which the actual observations conform to the design.</u>[④] In the network theory this quality of conformance is called reliability. The reliability of a network design is mainly based not upon the propagation of errors but rather on the degree to which the observations control each other. If they don't control each other within a network or a part of it there is no internal reliability, and a blunder in an observation will remain undetected and might falsify the result. The falsification of the network parameters due to bounders of any kind is described by the external reliability. To achieve a high reliability, networks have to be self-checking by means of independent redundant observations. Only then do we have sufficient information to apply statistical test, and can be quite sure that the observations conform with the assumptions made in the design computations. In summary, a check of reliability means a check of whether or not the adjustment model is correct. Computed reliability measures are numerical values for maximum possible amounts of those model errors which cannot be detected by the tests applied.

But the additional criterion, namely the economy of a design, is essential too. This is basically finding the answer to the question: "Given the set of observations to make, what is the least expensive way of making them?" From this point of view an extremely reliable design which needs a large number of redundant observations may not be justified. As a compromise it might be desirable to find those observations which are necessary to make the design sufficiently reliable, precise, sensitive and economical. Mathematically finding the best compromise is a problem of multi-criteria optimization. Because the target functions are nonlinear in the parameters and also discontinuous, it is not easy to formulate a strategy which can be proven to be optimal in all cases.

Words and Expressions

邹 optimization [ˌɒptəmaɪˈzeɪʃən] n. 最佳化；最优化
slide [slaɪd] n. 滑；滑动；幻灯片
subsidence [səbˈsaɪdns] n. 下沉；沉淀；陷没
evaluate [ɪˈvæljueɪt] vt. 评价；估计；求……的值
specification [ˌspesɪfɪˈkeɪʃn] n. 详述；规格；说明书；规范
defect [ˈdiːfekt] n. 过失；缺点
diminish [dɪˈmɪnɪʃ] v. 使减少；使变小
simulation [ˌsɪmjʊˈleɪʃn] n. 仿真；假装；模拟
iterative [ɪˈtərətɪv] adj. [数]迭代的；重复的；反复的
trial and error 累试法
optimum [ˈɒptɪməm] adj. 最适宜的
receipt [rɪˈsiːt] n. 收到；收条；收据
significantly [sɪɡˈnɪfɪkəntli] adv. 意味深长地；值得注目地
hinder [ˈhɪndər] v. 阻碍；打扰
standpoint [ˈstændpɔɪnt] n. 立场；观点
attune [əˈtjuːn] vt. 使相合；使合调
overambitious [ˌəʊvərˈæmbɪʃəs] adj. 野心太大的
exclude [ɪkˈskluːd] vt. 拒绝接纳；把……排除在外；排斥
swoop [swuːp] n. & v. 突然下降；猛扑
populate [ˈpɒpjuleɪt] v. 使居住；增殖；粒子数增加
singular [ˈsɪŋɡjələr] adj. 奇异的；单一的；非凡的；持异议的
non-singular 非奇异的
redundancy [rɪˈdʌndənsi] n. 冗余
criteria [kraɪˈtɪəriə] n. 标准；尺度
sensitivity [ˌsensəˈtɪvəti] n. 灵敏(度)；灵敏性

deformation [ˌdiːfɔːˈmeɪʃn] *n.* 变形
terminology [ˌtɜːmɪˈnɒlədʒi] *n.* 术语学
conformance [kənˈfɔːməns] *n.* 顺应；一致
fidelity [fɪˈdelɪti] *n.* 忠实；诚实；忠诚；保真度；逼真度；重现精度
falsify [ˈfɔːlsɪfaɪ] *v.* 伪造
falsification [ˌfɔːlsɪfɪˈkeɪʃn] *n.* 歪曲；弄虚作假；窜改；伪造
compromise [ˈkɒmprəmaɪz] *n.* 妥协；折中

☞ **Notes:**

① 建立工程测量控制网(测量网)，用于放样桥梁、隧道、水电站大坝和监测人工或自然建筑物如滑坡，区域沉降等。

② 权的优化设计问题是对给定的网形结构，找出相应的最优权或精度来确定观测方案，以满足控制网设计的要求。该问题原则上可以采用任何便利的平差程序，通过初始模拟、控制权或精度的变化和分析实际平差结果的相互影响来解决。

③ 这种方法的主要缺点是修改原有设计方案和接受修正方案预分析结果之间的时间延迟。结果处理缺乏时间上的连续性，明显降低了控制网优化设计的效率。

④ 或许对工程控制网(如监测网)更为重要的问题是一致性问题，即实际观测结果是否和设计结果相符合。

Unit 18　Construction Layout

Engineering surveys are conducted to obtain data essential for planning, estimating, locating, and layout for the various phases of construction activities or projects. After the necessary topographic maps are prepared and positions of the structures are established as well as the final plans for the project are available, the engineers, architects or building contractors need to know information about overall site grade and elevations to determine placement of site structures, or estimate the volume of dirt to be moved, and then to set the required horizontal and vertical positions for the structures.①

Construction surveying is the translation of construction plans into physical points on the ground that can be used as a basis for the actual construction. This type of surveying is sometimes called "construction layout" or "layout work". Obviously construction surveying is the inverse operation of the activities with which the surveyor is usually associated: the gathering of data regarding real points and positions on the ground and representing those positions on a map. The surveyor performing construction surveying is taking a "map" representing proposed structures and placing that information on the ground. As you can see, the surveyor is literally involved from start to finish of a construction project. The results of construction surveying are seen in almost any urban, suburban, and even rural setting. Almost any roadway, building, or other man-made structure probably had some amount of construction surveying involved. Construction surveying provides not only the horizontal location of new man-made structures, but also the vertical information required to ensure that surfaces drain or pipes flow as required.

Construction control: The first on-site job for the construction surveyor is to relocate the horizontal and vertical control used in the preliminary survey. Generally speaking, the horizontal and vertical control used in the preliminary survey may not be used again in construction surveying. Depending on the size and complexity of the project as well as the precision requirements, it may be necessary to reestablish the horizontal and vertical control in the areas of proposed construction. For example, the construction surveyor will establish one or more benchmarks in the general vicinity of the project. These benchmarks are placed away from the immediate vicinity of the buildings so they will not be destroyed by the construction operations, and are used to provide vertical control for the project. Once these are set, the surveyor will then establish a good many less permanent but more accessible benchmarks quite close to the project. The location of these less permanent points should be carefully selected so that turning points will ideally not be needed when elevations have to be set at the project.② Such careful selection of the points may result in critical time saving which is so important on construction project.

Layout: It is obviously critical for a construction job that the various parts of the structure be placed at the desired elevation. To accomplish this goal the construction surveyor will establish the reference lines or base lines before the actual layout measurements begins. For large construction projects, the usual procedure is to set a main base line down the centerline of the structure, such as the centerline of the bridge. The monuments are set along the centerline at each end beyond the area of the construction work. The monuments along the ends of the line may be occupied by the surveyor and will enable him to check and reset points within the construction areas. The stakes and points set during the process of construction surveying are not usually set at the actual construction point, but usually on some sort of offset. This is done so that the survey stake is not disturbed by excavation or other activities that will take place at the actual point of construction. The stakes are marked with an "offset" and a "grade", which provides the construction crew with the special relationship of the construction stake to the actual point of construction.③ The "offset" is the distance from the survey stake to the horizontal position of the actual point of construction, and would typically be three feet for curb and gutter to ten feet for underground pipes.

Techniques of layout: In construction layout measurement, the data necessary to establish the direction and distance from a control point to locate a construction point can be entered into the instrument via the keyboard or directly from an office computer. Then the surveyor guides the person holding the prism along the line of computed direction until the distance to the point to be located agrees with the computed distance. Various techniques may be applied to accomplish this goal. A very popular technique called free station permits the surveyor to set up the total station at any convenient position and then to determine the coordinates and elevation of that instrument position by sighting previously coordinated reference stations.④ After the instrument has been set up over this instrument position (a control point) and properly oriented, angles or azimuths from the control point and distances to each layout point may be indicated. Now many total stations have such functions that the coordinates and elevations of the layout points may be uploaded into the total station, the instrument's display shows the left/right, forward/back, and up/down movements needed to place the prism in each of the desired positions.

As-built surveys: When the construction surveying has been completed, a final survey is performed. The final survey includes cross sections and locations that are used for final payments to the contractor and for the completion of an as-built drawing. These are the as-built surveys and they are used to check the contractor's work and show locations of structures and their components (water line, sewer, etc.), which will be needed for future maintenance, changes, and new construction.

Words and Expressions

architect [ˈɑːrkɪtekt] n.建筑师
literally [ˈlɪtərəli] adv.照字面意义；逐字地
urban [ˈɜːrbən] adj.城市的；市内的
suburban [sʌˈbɜːbən] adj.郊外的；偏远的
drain [dreɪn] n.排水沟；消耗；排水
on-site 场区内；厂区内；工地上
complexity [kəmˈpleksɪti] n.复杂(性)；复杂的事物
reestablish [ˌriːɪˈstæblɪʃ] v.重建；使复原；使复位
vicinity [vɪˈsɪnɪti] n.邻近；附近；接近
excavation [ˌekskəˈveɪʃn] n.挖掘；发掘；挖掘成的洞；出土文物
crew [kruː] n.全体人员；(工作)队
curb [kɜːrb] n.路边；人行道边缘
gutter [ˈɡʌtər] n.水槽；檐槽；排水沟；槽；贫民区
payment [ˈpeɪmənt] n.付款；支付；报酬；偿还；报应；惩罚
sewer [ˈsuːər] n.下水道；缝具；缝纫者

☞ Notes：

① 工程师、建筑师或建筑承包商需要了解整个工地的坡度和高程来进行施工现场建筑物的布置，或估算需要开挖的土方量，然后放样建筑物的水平位置和高程。
② 这些基准点应埋设在远离建筑物的地方，以免在施工过程中被损毁，并为工程提供高程控制。一旦这些基准建立之后，测量员就可以在紧邻工程施工处外建立大量便于使用的非永久性基准点。这些非永久性基准点的位置需要细心地选择，以便在理想条件下工程施工放样中不需要设置转点就可以放样高程。
③ 木桩上标注有"偏差"和"坡度"，为施工队提供了木桩到实际施工点的相应关系。
④ 一种非常常用的技术称为自由设站法，这种方法允许测量员在任何便利位置架设全站仪，然后通过联测以前的已知坐标参考点，来求出全站仪设站点的坐标和高程。

Terms Highlights

engineering survey 工程测量
engineering control network 工程控制网
construction control network 施工控制网
control network for deformation observation 变形观测控制网
precision 精度
sensitivity 灵敏度

reliability 可靠性
construction survey 施工测量
free station 自由设站法
as-built survey 竣工测量
setting-out survey, construction layout 施工放样
alignments survey 定线测量
public engineering survey 市政工程测量
building engineering survey 建筑工程测量
road engineering survey 道路工程测量
hydrographic engineering survey 水利工程测量
bridge survey 桥梁测量
tunnel survey 隧道测量
surveying for site selection 厂址测量
survey of present state at industrial site 工厂现状图测量
setting-out of main axis 主轴线放样
building axis survey 建筑轴线测量
property line survey 建筑红线测量
construction plan 施工平面图
site map 工地(总)平面图
cross section 横断面；断面图；剖面图
profile diagram, profile 纵断面图
profile survey 纵断面测量
cross-section survey 横断面测量

Unit 19　Tunnel Surveying and Construction Methods

According to Wordtips, a tunnel is an artificial underground passage, especially one built through a hill or under a building, road or river while it is described as a passage built underground, for example, to allow a road or railway to go through a hill, under a river, etc., after *Oxford Learner's Dictionaries*. A tunnel is an underground or underwater passageway by digging through the surrounding soil, earth, rock and enclosed except for its entrance and exit commonly at each end (Wikipedia). In general, a tunnel is characterized to be relatively long and narrow with its length often much greater than twice the diameter and completely enclosed all around, but at the two end portals.

There are different types of tunnels. A tunnel may function as pedestrian crosswalk, part of a vehicular road or railway, waterway for a canal, passageway for mining operation, or any other similar utilities (water supply or drainage). More and more tunnels are built as part of municipal transportation systems in modern cities to preserve open space above the ground for enhancing transportation safety, visual environment and people's recreation activities. Tunnels also provide easy passageways to underground structures, which can sustain earthquakes and other natural disasters, consume less energy for heating and cooling, do not need attractive exterior finishes, and require low maintenance.

Although a subway means an underground rapid transit system in Canada and the United States of America, a pedestrian tunnel or an underpass beneath a road for humans and even wildlife to safely cross to the other side of the road is usually called an underpass or a subway in the United Kingdom and other Commonwealth countries.

Guiding a tunnel boring machine following the designed line and grade is a complex task, with which a surveyor or surveying engineer is normally charged responsibly. Accordingly, a well-established geodetic control network is essential. To establish a survey control system for an interconnected network of tunnels, further complexity is added to the course of surveying. While a tunnel is being constructed by multiple tunnel boring machines at the same time, might even by different construction contractors, the surveying crew must assure that they are interconnected with each other within the allowable breakthrough tolerance for both of the axial line and grade.① To achieve such a degree of specific accuracy requirements, it is important for a project manager to develop a high quality survey program by taking into account all the steps in the sequence of surveying operations from design and planning stages before the construction to the end of the construction staking-out. This surveying program must assure that the construction of a tunnel satisfies the specified breakthrough error tolerance and must be consistent with the survey methods used by the contractors.

Tunnel Surveying

Tunnel surveying refers to the surveying work during the design, construction, completion, and operation stage of a tunnel project. The cost of surveying work is relatively small in comparison with the expenditure on the entire tunnel construction. Nevertheless, surveying involvement plays its essential role before the construction, during the tunnel construction, operation and maintenance. The monetary loss may be significant in case any faulty surveying result occurs with construction operations.

First, before the start of construction, the surveying engineers take on preliminary surveys to map the project site, and establish primary surface control networks in order to serve for surface construction and tunnel guidance system underground. Then, during the construction, the surveying engineers are in charge of conducting the survey work associated with the tunnel construction and coordinating its breakthrough process at the interfaces between individual construction contractors.[②] The surveying engineers thus assure that the working faces of a tunnel, guided by surveying systems, from any two adjacent contractors underground can break through agreeably at their interconnecting interface and the reliable monitoring of surface settlement as well, which partially ensure the safety of tunnel construction.

Therefore, according to the sequence of surveying operations in tunnel construction, the associated surveying work may be divided into the following three parts: pre-construction survey, construction survey and post-construction survey, which can be summarized as the following specific tasks:

(1) Preliminary survey overall serves tunnel alignment design, determination of clearances to the existing structures and topographic features and locations of existing underground utilities as well.

(2) Control survey consists of surface control surveys to serve as the basis of surveying control for the entire tunnel from design to construction; correlation surveys to transfer the coordinates, azimuths and elevations from the surface control points to underground; underground control surveys to guide and control the tunnel excavation.

(3) Tunneling guidance functions to monitor and control the position and attitude of a tunneling machine throughout the tunnel excavation for drilling and lining.

(4) Breakthrough survey is to determine the deviation of the actual alignment from the designed central line.

(5) Wriggle survey determines as-built clearances and provides a permanent record

of the as-built tunnel.

(6) Deformation monitoring detects possible deformation of the ground, structural components, buildings and utilities during and after construction in order to ensure the site safety.

Tunnel Construction Methods

The process to build a tunnel, i.e., the tunnel construction, is commonly called excavation. Tunnel construction methods are differentiated between blasting and mechanized tunneling, or excavation methods. <u>In order to select a technically feasible and economical construction method, the main aspects that need to be taken into consideration are: tunnel length, tunnel cross section, geological, hydro-geological and geotechnical conditions, tunnel lining, construction logistic, health and safety, and construction time.</u>③

(1) **Drill & Blast (D & B) Method**: As the name suggests, the procedure using the drill and blast excavation method is as follows:

- Drill a number of blast holes into the rock, which are then loaded with explosives.
- Detonate the explosives to break out the rocks.
- Remove muck, a process known as mucking, and apply tunnel support in the form of shotcrete, anchors, lattice arches and reinforcement mats, etc.
- Repeat the process until the desired excavation is achieved.

The positions and depths of the holes (and the explosion load each hole receives) are determined by a carefully constructed pattern, which, together with the correct timing of each explosion, will guarantee that the tunnel will have an approximately circular cross-section.

(2) **Tunnel Boring Machine Method**: A tunnel boring machine (TBM), also known as a "mole", is a machine used to excavate tunnels with a circular cross section through a variety of soil and rock strata.

As an alternative to the D & B methods, TBMs are increasingly used to excavate tunnels with a circular cross section through a variety of subterranean matters; hard rocks, sands or almost anything in between. As a TBM moves forward, the round cutter heads cut into the tunnel face and splits off large chunks of rock. The cutter head carves a smooth round hole through the rock, the exact shape of a tunnel. The conveyor belts carry the rock shavings came out from the TBM to a dumpster in the back of the machine. Tunnel lining is the wall of the tunnel and consists of precast concrete segments to form rings. Tunneling with TBMs is much more efficient and can result in shortened completion times indeed.

(3) **Shield Tunneling Method**: A tunneling shield is a protective structure used during the excavation of tunnels. When excavating through soft ground, liquid, or

unstable ground, there is a high risk of a pit accident to the workers and the project itself, because it is difficult to prevent mud and water from seeping in and collapsing the tunnel heading. A tunneling shield serves as a temporary support structure and is usually put in place for the excavated tunnel section for a short time period until the excavated section can be lined with a permanent support structure made up of bricks, concrete, cast iron, or steel.

Nowadays shield tunneling method usually refers to the tunnel construction using modern tunnel boring machines.

(4) **Cut-and-Cover Method**: Cut-and-cover is a simple tunnel construction method for shallow tunnels and also belongs to the oldest tunneling method. It involves digging of a trench, tunnel construction, and roofing over with an overhead support system strong enough to carry the load and returning the surface above the tunnel to its original state.

Shallow tunnels are often built by using the cut-and-cover methods (if under water, of the immersed-tube type). While deep tunnels are excavated, a tunnelling shield is often used. For intermediate levels, both methods are possible. The major disadvantage with the cut-and-cover method is the widespread disruption generated at the surface level during the construction.

(5) **Immersed-Tube Method**: Pioneered by the American engineer W. J. Wilgus in 1903, the immersed tube (also called Sunken Tube) tunnel technique is specifically used for underwater tunneling to cross a body of shallow water. This technique uses hollow box sectioned tunnel elements. <u>They have been prefabricated in reinforced concrete, are moved into the harbor, and placed into a trench pre-dredged in the harbor bed. Once the elements are in position, they are joined together to form a tunnel.</u>④ The trench is then refilled so that the harbor bed is returned to its original level.

Words and Expressions

portal [ˈpɔːrtl] n. 门；入口
pedestrian [pəˈdestriən] n. 步行者；行人
sustain [səˈsteɪn] v. 支持；支撑；忍受；经受
commonwealth [ˈkɒmənwelθ] n. 国家；独立的政治共同体；英联邦
tolerance [ˈtɒlərəns] n. 公差；容限；限差
interface [ˈɪntəfeɪs] n. 接合点；结合点；边缘区域；界面；接口
expenditure [ɪkˈspendɪtʃər] n. 花费；支出；开支
alignment [əˈlaɪnmənt] n. 排成直线；路线
attitude [ˈætɪtjuːd] n. 姿态；姿势；态度；看法
lining [ˈlaɪnɪŋ] n. 衬里；内胆；里子
wriggle [ˈrɪgl] v.& n. 扭动；蠕动
geotechnical [ˌdʒiːəʊˈteknɪkl] adj. 岩土工程技术的

detonate [dɪ'təʊneɪt] v.爆炸；引爆
shotcrete [ˈʃɒtkriːt] n.压力喷浆；喷射的水泥砂浆
anchor [ˈæŋkər] n.锚；顶梁柱；刹车
lattice [ˈlætɪs] n.花格结构；格子木架；交错结构
reinforcement [ˌriːɪnˈfɔːsmənt] n.加固物；增强材料
subterranean [ˌsʌbtəˈreɪnɪən] adj.地面下的；在地下进行的
dumpster [ˈdʌmpstər] n.大垃圾桶
seep [siːp] v.渗出；渗漏
cast iron [ˈkɑːstˈaɪərn] n.铸铁；生铁
immerse [ɪˈmɜːrs] v.浸；投入；陷入；沉浸于
sunken [ˈsʌŋkən] adj.沉没的；没入水中的；凹陷的；下陷的；深陷的
prefabricate [ˈpriːˈfæbrɪkeɪt] v.预先建造；预制
pre-dredge [priːˈdrɛdʒ] v.疏浚；挖掘

☞ Notes：

①当一个隧道由多台隧道掘进机同时施工，甚至可能由不同的施工承包商施工时，测量人员必须确保开挖隧道中线和坡度在允许的贯通限差范围内相互衔接。
②在施工开始前，测量工程师首先要进行初测，测绘工程选址区的地形图，建立首级地面控制网为地面施工和地下隧道导向系统服务。然后，在施工期间，测量工程师负责隧道施工相关的测量工作，并协调不同承包商在隧道衔接处的贯通。
③为了选择技术上可行且经济的施工方法，需要考虑的主要因素有：隧道长度、隧道横截面、地质、水文地质和岩土条件、隧道衬砌、施工物流、健康和安全以及施工时间。
④将预制钢筋混凝土管段移至港口，沉放入预先挖掘好的海底沟槽内，一旦管段就位，就可以将相邻管段连通形成隧道。

Terms Highlights

breakthrough survey 贯通测量
breakthrough tolerance 贯通限差
breakthrough error 贯通误差
surface control survey 地面控制测量
correlation survey 联系测量
underground control survey 地下控制测量
construction traverse 施工导线
basic traverse 基本导线
primary traverse 主要导线
tunnel guidance system 隧道引导系统
preliminary survey 初测
wriggle survey 收边测量；断面测量

tunnel construction method 隧道施工方法
drill and blast excavation method 钻爆开挖法
tunnel boring machine (TBM) 隧道掘进机
shield tunneling method 盾构法
cut-and-cover method 明挖法
immersed tube (sunken tube) tunnel 沉管法
tunnel lining 隧道衬砌
vertical shaft 竖井

Unit 20 Deformation Monitoring of Engineering Structure

Overview

Deformation refers to the changes of a deformable body (natural or man-made objects) undergoes in its shapes, dimension and position in space and time domain. Due to factors such as changes of ground water level, tidal phenomena, tectonic phenomena, etc., engineering structures (such as dams, bridges, high rise buildings, etc.) are subject to deformation. Deformation of engineering structures is often measured in order to ensure that the structure is exhibiting a safe deformation behavior. Cost is more than offset by savings and by improvements in safety both during and after constructions.① Expanded resource development, the trend towards potentially-deformation-sensitivity engineering and construction projects, and growing geosciencetific interest in the study of crustal movement have all combined to increase awareness of the need for a comprehensive integrated approach to the design and analysis of such deformation surveys. Therefore it is important to measure this movements for the purpose of safety assessment and as well as preventing any disaster in the future.

Deformation Monitoring of natural and man-made structures is an engineering survey activity during which repeated observations are made within a specified time frame for the purpose of detecting and quantifying movements of structures.② Such monitoring could be of a routine nature (e.g., a dam at high and low water marks) or made necessary by an abnormal condition (e.g., major works near a tall building). Under ordinary circumstances, the interval of time between monitoring and analysis may extend over several days or more. Under critical condition, this may have to be nearly instantaneous in order to provide a warning, if necessary. The volume of data may consist of only several items, in the simplest routine investigation, or of hundreds or thousands of different data, in very complex or critical conditions. The rate of monitoring may be annually, monthly, weekly, daily, hourly, or even more frequently. Detecting and quantifying movements require the use of very precise equipment. Such movements are very small and to accurately measure them requires meticulous fieldwork as well as rigorous analysis of observed data. Deformation monitoring techniques can be generally divided into geotechnical, structural and geodetic (survey) methods. The geodetic methods (highly understood by engineering surveyors) that can be used are Global Positioning System

(GPS), close-range photogrammetry with the use of terrestrial camera, precise theodolite and levels, total station, laser scanners, and vibration monitoring systems, a very long baseline interferometry and satellite laser ranging. A number of traditional surveying techniques have also been modified and applied to yield the highest possible accuracies. Geodetic measurements involve the observables of horizontal angles, or direction; spatial distance; and height difference. Geotechnical measurements consist of similar geometric quantities, but over a much smaller extent than for geodetic measurements (distances to a few meters rather than hundreds of meters), as well as measurements of the physical or mechanical state of the object being monitored.

The survey methods can be further subdivided into the survey network method and direct measurement method. In geodetic method there are two types of geodetic networks, namely the reference (absolute) and relative network.

Monitoring Schemes

A deformation survey requires the assessment of project expectations. This would include accuracy statements in order to detect the movement and external effects of the object suspected of movement. The observation period and frequency must also be established. The selection of most appropriate technique or combination of techniques for any particular application will depend upon cost, the accuracy required, and the scale of the survey involved. Therefore several aspects related to the optimal design of the networks, measurement and analysis techniques suited to the monitoring surveys have to be considered. The design of monitoring scheme should satisfy not only the best geometrical strength of the network but should primarily fulfill the needs of subsequent physical interpretation of the monitoring results. Selection of monitoring techniques depends heavily on the type, the magnitude and the rate of the deformation. Therefore, the proposed measuring scheme should be based on the best possible combination of all available measuring instrumentation. A common feature for both geodetic and satellite methods in monitoring scheme involves the following three stages:

(1) The design of the scheme required to undertake the deformation survey must consider the instrumentation to be used, the geometry of the network, the location of observables points, the types of observables, preanalysis of possibly environmental influences, and frequency of observation of the observables to the expected form of deformation. The reference datum must be appropriate, secure, and stable and not influenced by the suspected or anticipated movement within the local site area.

(2) The execution process that runs a designed network into reality should be carefully done, which deals with both the documentation of the proposed network stations and the actual field measurement techniques in which the equipment used must be adjusted and in good working order with appropriate

calibrations completed.③

(3) Data processing deals with the processing and analysis of the collected geodetic data, quality assurance and control that must be followed to confirm the expected results. Processing occurs simultaneously with capture and subsequent to capture with different tasks at each time. During capture, the points involved must be identified an ancillary observations, such as temperature, must be requested along with dealing with the observation. The observation would likely be repeated in order to obtain a mean and an estimate of its standard deviation and the mean would be compared with the predicted or most recent value as a check on consistency.④ Subsequent processing would further reduce the observation and the data into series file or campaign file following the structure of the data management system. Consistency would be checked either again for the observation or further for the reduced data.

Trend Analysis

Once more than one campaign has been observed or once enough data are contained in a series, it is necessary to determine the tendency that is being exhibited in space or over time or both. The observed tendencies are then brought together to suggest possible forms of models, i.e., the choice of parameters to be estimated. The trend analysis acts as a filter by extracting the behavior of interest, e.g., the annual trend or rate, from the time series, e.g., the noise being the seasonal cycle that is a reaction to the change in temperature. The extracted trends become the input or "observations" in the modeling and, therefore, it is necessary to have measures of variance associated with each trend. The spatial trend can be derived for one or two dimensional networks by considering the differences in coordinates estimated in the individual campaign adjustments, i.e., comparing campaigns. Spatial trend can also be derived vertically for subsidence profiles.⑤

Finally the reporting of the deformation survey should include the history of the project as some projects could extend over a significant period of time. The report should also include the methodology containing results of the individual campaign (epoch), the geometrical displacements and the quality of the displacements, graphic depiction and recommendations or conclusions.

Words and Expressions

domain [dəʊˈmeɪn] n. 领域；领土；领地；(活动、学问等的)范围
tectonic [tekˈtɔnik] adj. [建]构造的；建筑的
awareness [əˈweənes] n. 知道；晓得
assessment [əˈsesmənt] n. 估价；被估定的金额

disaster [dɪˈzɑːstər] n.灾难；天灾；灾祸
abnormal [æbˈnɔːməl] adj.反常的；变态的
instantaneous [ˌɪnstənˈteɪniəs] adj.瞬间的；即刻的；即时的
investigation [ɪnˌvestɪˈɡeɪʃn] n.调查；研究
meticulous [məˈtɪkjələs] adj.小心翼翼的；一丝不苟的
vibration [vaɪˈbreɪʃn] n.振动；颤动；摇动；摆动
mechanical [məˈkænɪkl] adj.机械的；机械制的；机械似的；呆板的
interpretation [ɪnˌtɜːprɪˈteɪʃn] n.解释；阐明；口译；通译
instrumentation [ˌɪnstrəmenˈteɪʃn] n.使用仪器
suspect [səˈspekt] v.怀疑；猜想；对……有所觉察
anticipated [ænˈtɪsɪpeɪtɪd] adj.预先的；预期的
ancillary [ænˈsɪləri] adj.补助的；副的
campaign [kæmˈpeɪn] n.[军]战役；(政治或商业性)活动；竞选运动
filter [ˈfɪltər] n.滤波器；过滤器 v.筛选；过滤
epoch [ˈiːpɒk] n.新纪元；时代；时期；时间上的一点；[地质]世
recommendation [ˌrekəmenˈdeɪʃn] n.推荐；介绍(信)；劝告；建议

☞ Notes：

① 为了确保工程建筑物处于安全的变形范围内，我们需要经常地进行建筑物的变形观测。建筑物建成前后的安全保障和改善足以弥补变形观测所花的费用。
② 对各种人工或自然建筑物进行变形观测是工程测量的一项工作，为了监测和量化建筑物的变形大小，需要在特定的时间范围内进行周期性的重复观测。
③ 实际布网观测的过程必须认真细致，该过程要考虑到设计网点的资料信息，以及所采用的实际野外观测技术，使用的仪器设备要经过调试校准，并处于良好的工作状态。
④ 观测一般要有重复观测，以获得该观测值的均值和标准差估值，均值应用来比较预测值或近期观测值以检查数据的一致性。
⑤ 对每个周期的观测值进行平差，并比较不同周期坐标估值的差值，可以描述一维或二维监测网的空间变化趋势，还可以得到沉降剖面的垂直变化趋势。

Terms Highlights

deformation monitoring(observation) 变形监测(观测)
displacement observation 位移观测
settlement (subsidence) observation 沉陷观测
fissure observation 裂缝观测
deflection observation 挠度观测
oblique observation, tilt observation 倾斜观测
mining subsidence observation 开采沉陷观测
observation of slope stability 边坡稳定性观测

method of tension wire alignment 引张线法
collimation line method 视准线法
method of laser alignment 激光准直法
minor angle method 小角度法
direct plummet observation 正垂线观测
inverse plummet observation 倒垂线观测

Part II

Advanced Technologies in Geomatics Engineering

Unit 21　GNSS Fundamentals

What Is GNSS?

GNSS stands for Global Navigation Satellite System, which is the collective term for any global navigation satellite systems that provide users with Positioning, Navigation and Timing (PNT) as a combination of three distinct and constituent capabilities as stated at www.transportation.gov/pnt/:

- ***Positioning***, *the ability to accurately and precisely determine one's location and orientation two-dimensionally (or three-dimensionally when required) referenced to a standard geodetic system (such as World Geodetic System* 1984, *or WGS*84);
- ***Navigation***, *the ability to determine current and desired position (relative or absolute) and apply corrections to course, orientation, and speed to attain a desired position anywhere around the world, from sub-surface to surface and from surface to space; and*
- ***Timing***, *the ability to acquire and maintain accurate and precise time from a standard (Coordinated Universal Time, or UTC), anywhere in the world and within user-defined timeliness parameters. Timing also includes time transfer.*

Up to now, four global navigation satellite systems have been developed: Global Positioning System (GPS) owned and operated by the United States, the GLObal NAvigation Satellite System (GLONASS) owned and operated by the Russian Federation, the GALILEO system owned and operated by the European Union, and the BeiDou Navigation Satellite System (BDS) owned and operated by the People's Republic of China. A GNSS provides all-time, all-weather and continuous high-accuracy positioning, navigation, timing services worldwide.

A GNSS consists of three segments: the space, control, and user segments. Although each GNSS functions independently, users may simultaneously receive the signals from the satellites belonging to more than one GNSS system.

Space Segment: GNSS Constellation

The space segment of a GNSS consists of satellites, which construct a GNSS constellation. The satellites of GPS, GLONASS, Galileo and BDS are arranged into a number of medium Earth orbits (MEOs) continuously orbiting the Earth with the capability of emitting their specific signals that can be tracked by user's GNSS equipment, and also being monitored and maintained by their own control segments. A fully operational constellation normally comprises at least 24 satellites. For example, <u>a nominal GPS constellation by design comprises 24 satellites orbiting the Earth with a period of about 12 hours at an altitude of about 20,200 km in six equally-spaced orbits with four satellites on each. The orbits are tilted to the Earth's equator by 55 degrees and separated 60 degrees right ascension of the ascending node.</u>①

Control Segment: Ground Segment

The control segment, also referred to ground segment or operational control system, is responsible for maintaining the proper operation of satellites. With the GPS, for instance, its control segment consists of a global network of ground facilities that track the satellites, monitor their transmissions, perform analysis, and send commands and data up to the constellation. Specifically, the current GPS control segment includes a master station and an alternate backup master control station, 11 command and control antennas, and 16 monitoring sites. The monitoring stations measure ranges from the satellites and send them to the control stations. The master control station processes the measurements to estimate satellite orbits (ephemerides) and clock errors, among other parameters, and to generate the navigation message. These corrections and navigation message are uploaded to the satellites by the Master Control Station. The master control station is fully backed up by the alternate master control station.

User Segment: GNSS Users

The user segment of a GNSS consists of all the GNSS receiver equipment, which tracks the signals sent from its satellites and makes the measurements and ephemeris available to estimate user's 3D positions and time. A user equipment basically consists of a GNSS antenna and a receiver with built-in microcomputer, which is also quite commonly called a GNSS receiver. Specifically, the GPS user segment consists of L-band radio receivers/processors and antennas, which receive the GPS signals, deliver pseudoranges and carrier phase measurements together with other information, and solve the navigation equations in order to obtain their coordinates and provide a very accurate time. There are many different types of the GNSS receivers, which can be classified in different ways. One may categorizes them after their usages: navigation, accurate positioning or timing, and surveying, etc., in most important economic activities for civilian users, or the military GNSS navigation, reconnaissance, and missile guidance systems.

GNSS Positioning

The basic function of all GNSS systems is to use a GNSS receiver to determine its own position by applying a technique called satellite ranging after trilateration, which involves measuring distances from the GNSS receiver to at least four GNSS satellites being tracked by it at the same time. GNSS satellites broadcast the radio signals using their carrier waves, on which the ranging codes and the navigation message (i.e., the ephemeris) with their status, and corrections for on-board atomic clocks and ionospheric delay etc., are modulated. <u>In general, GNSS devices track, record the radio signals sent by GNSS satellites, and further decode the modulated data to deliver the pseudoranges and carrier phase measurements along with the forecasted satellite ephemeris etc., from which either the single point positioning or other relative, even network positioning can be conducted after the principle of trilateration.</u>[②]

In positioning based on GNSS, all the satellites geometrically play the same role as the control monuments in the traditional geodetic control networks except they are continually moving on their own orbits. The position of each satellite is practically computed in the form of the 3D Earth Centered Earth Fixed (ECEF) coordinates for a given time instant through its ephemeris. By taking the GPS as an example, there are different types of ephemeris: (1) Almanac, which consists of coarse orbit and status information of each satellite, an ionospheric model, and information to relate the GPS derived time to the UTC time. (2) Ephemeris, also called the broadcast ephemeris, which is the major part of the navigation messages modulated on GPS signal, transmitted by GPS satellites and recorded by GPS receivers. It is used to estimate current satellite's location and predict their future location. (3) Precise ephemeris, also called GPS precise ephemeris, or GPS orbit product, which is the post-processed ephemeris made freely available to the general public by IGS (International GNSS Service, igs.org/products/).

GNSS Positioning Techniques

Since 1980s, GNSS technology has been continually developed and innovated for civilian applications. Now, it becomes the most popular technology to fulfill land, airborne and marine positioning tasks either in static or kinematic mode. There are different GNSS positioning techniques developed for different applications which can be classified in four different ways: i) absolute (or single point) or relative positioning, ii) static or kinematic positioning, iii) real-time and post-processing positioning, iv) the positioning using pseudoranges or both of pseudoranges and carrier phases. Depending on the applications and accuracy requirements, one can decide for the best suitable positioning technique. A brief summary of different GNSS positioning techniques is given below.

(1) **Absolute positioning**

Absolute positioning means that the position of a receiver is determined directly in relation to the GNSS satellites. As shown in Figure 1, one simply employs a single GNSS receiver to fulfill the positioning task by using satellite ranging, which utilizing the distance measurements between a GNSS receiver and the locked GNSS satellites. Hence, the corresponding positioning mode is simply called single-point positioning.

Figure 1　GNSS Single-point Positioning

Specifically with GPS, for example, the two positioning services by design: SPS (Standard Positioning Service) and PPS (Precise Positioning Service) belong to this category. The former utilizes the pseudoranges (L1 C/A) while the latter utilizes the military precise codes (P1 and P2 codes) in single point positioning. The solution performance of the pseudorange only point positioning technique can be improved by ***carrier-phase smoothed pseudorange method***.

The reachable accuracy in single-point positioning depends on the equipment, the observation time, the satellites' geometry in the sky, the measurement model, and the software used. The global average position accuracy in Standard Point Positioning is better than 8.0m (95% horizontal error) and 13.0m (95% vertical error) while the global average velocity accuracy is better than 0.2m/s (95% velocity error, any axis) and the time transfer accuracy is better than 30ns (95% of time, Space-in-Space only) [Office of DoD, 2020].

Under this category, **Precise Point Positioning** (PPP) belongs to the most recent technique, which removes or models GNSS system error to provide a high level of position accuracy from a single receiver. The PPP was seriously studied in 1990s and extensively progressed during the last two decades using dual frequency measurements and the widely available corrections from GNSS networks. It has been demonstrated that cm level PPP solution is achievable in post-processed, static positioning mode and potentially for kinematic applications [Rizos et al., 2012].

All of the above mentioned single point positioning techniques may be applied in static or kinematic positioning, real-time as well as post-processing, even integrated with other sensors such as Inertial Measurement Units (IMUs).

(2) **Relative positioning**

In relative positioning, the position of a GNSS receiver is determined relative to one or more GNSS receivers set up at other nearby control stations. Figure 2 outlines the

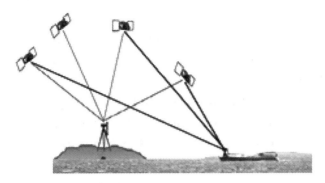

Figure 2　Relative GNSS Positioning

basic concept how relative GNSS positioning works, wherein the position of the receiver on the right is determined relative to the position of the receiver on the left. Relative positioning can be completed by applying multiple variations of similar GNSS positioning techniques. First, differential GNSS (DGNSS) technique can be deployed, which is realized by providing users with range corrections to the GNSS signals, commonly the pseudoranges or with the positional corrections to the derived user's positions from user's receivers toward significant improvement of positional accuracy. For very large working area, a DGNSS network consisted of multiple permanent GNSS reference stations is regionally or continentally established so that a geostationary satellite may be rent to upload the DGNSS corrections to serve the subscribed users in real time for both of the static and kinematic applications.[3] Here, both of the GNSS pseudorange and carrier phase measurements are utilized mostly in the form of double differenced measurements between two simultaneously operated receivers so that most of the systematic errors can be greatly reduced, even cancelled out. Besides, if one receiver is set up at a fixed base station, the second receiver can be a rover mounted on a moving platform to determine a 3D trajectory. In case, if the receivers are capable of processing carrier phase measurement in real time, the rover receiver is able to provide up to cm level accuracy of kinematic positioning solution, which is as known as RTK (Real-Time Kinematic). In case, multiple base stations are set up, a RTK GNSS network is established, which is summarized separately right below.

(3) **Network RTK**

The single baseline RTK was developed in the mid-1990s to utilize the carrier phase measurements to enhance the solution performance. In general, a reference receiver is set up at a base station for transmitting its raw measurements or corrections to a rover receiver through a specific radio link. The data processing at the rover site includes ambiguity resolution of the differenced carrier phase data and coordinate estimation of the rover position. The single baseline RTK technique has been proven to be a cm-level accuracy of real time positioning technique. One significant drawback of this single base RTK approach is that the maximum distance between reference and rover receiver must not exceed 10 to 20 kilometers in order to be able to rapidly and reliably resolve the carrier phase ambiguities.

The introduction of Network RTK is to purposely achieve: a significant reduction of the number of the GNSS base stations; a large coverage of RTK service and a long term local and regional geospatial infrastructure.④ A significant number of RTK Networks have been established worldwide during the last two decades, of which most of them provide multi-purpose services. Some of them function as part of the authoritative geospatial infrastructure while the others are commercial GNSS Networks owned by private sectors. The German Satellite Positioning Service (SAPOS), SmartNet (Leica), TopNet (Topcon) and Can-Net are the examples of GNSS RTK networks.

The current trend in advancing the usage of GNSS technology is to combine PPP and RTK towards emerging applications such as autonomous vehicles and unmanned logistics as it has several advantages including high precision and hign reliability, full flexibility, and good privacy.

Words and Expressions

collective [kəˈlektɪv] adj.聚集而成的；集体的；共同的；集合的
simultaneously [ˌsɪməlˈteɪnɪəslɪ] adv.同时地
constellation [ˌkɒnstəˈleɪʃn] n.[天]星座；星群
altitude [ˈæltɪtjuːd] n.高度；海拔；高地
equally-spaced [ˈiːkwəli ˈspeɪst] adj.等距的；间距相等的；同样间隔的
tilt [tɪlt] v.倾斜；侧倾 n.倾斜；侧倾；倾向；偏向；倾斜面；斜坡；俯仰转动
ascension [əˈsenʃn] n.上升(right ascension 赤经)
ascend [əˈsend] v.登上；攀登；升起；上升(ascending node 升交点)
antenna [ænˈtenə] n.天线
ephemeride [ɪˈfemərɪd] n.[天]星历表；历书
upload [ˈʌpˌləʊd] v. & n.上载
pseudorange [ˈsjuːdəʊrəndʒ] n.伪距
carrier [ˈkærɪər] n.搬运者；载体
reconnaissance [rɪˈkɒnɪsns] n.勘测；侦察；搜索
decode [diːˈkəʊd] v.译；解读；转换
almanac [ˈɔːlmənæk] n.年历；历书；年鉴
coarse [kɔːs] adj.粗的；粗糙的
ionosphere [aɪˈɒnəˌsfɪər] n.电离层
static [ˈstætɪk] adj.静的；静态的；静止的；不动的
kinematic [ˌkɪnɪˈmætɪk] adj.非静止的；运动学的；运动学上的
differential [ˌdɪfeˈrenʃl] adj.差分的；差动的；微分的
deploy [dɪˈplɔɪ] v.部署；利用
geostationary [ˌdʒiːəʊˈsteɪʃ(ə)neri] adj.与地球旋转同步的；与地球同步的

subscribe [səbˈskraɪb] v.订阅；订购；认购；支付
rover [ˈrəʊvər] n.流浪者；漫游者；自由队员
ambiguity [æmbɪˈɡjuːəti] n.意义不明确；含糊
autonomous [ɔːˈtɒnəməs] adj.自治的；独立自主的；自律的；自制的
logistics [ləˈdʒɪstɪks] n.(pl.) 后勤；物流；货物配送

☞ **Notes:**

①GPS 系统设计的星座由 24 颗卫星组成，它们分布在 6 个等间距轨道上，每个轨道有 4 颗卫星，卫星轨道高度为大约 20200km，绕地球运行的周期约 12 小时。6 个轨道平面相对于赤道的倾斜角为 55 度，相邻轨道的升交点赤经之差为 60 度。
②通常，采用 GNSS 接收设备跟踪、记录由 GNSS 卫星发送的无线电信号，并进一步通过调制数据的解码来获得伪距和载波相位测量值以及卫星的广播（或预测）星历等。从而，按照三维测边网测量原理，进行单点或相对定位，甚至是网络定位。
③对于大面积的工作区域，可由多个永久性 GNSS 参考站建立区域或洲际 DGNSS 网，租用商业地球同步卫星上传差分改正数据，为付费用户的静态和动态定位应用提供实时服务。
④引入 RTK 网的目的是：显著减少 GNSS 基站数量，加大 RTK 服务范围以及为当地或地区建设长期的地理空间基础设施。

Terms Highlights

Global Navigational Satellite System (GNSS) 全球导航卫星系统
Global Positioning System (GPS) GPS 定位系统[美]
GLObal NAvigation Satellite System (GLONASS) GLONASS 导航卫星系统[俄]
GALILEO 伽利略系统[欧盟]
BeiDou Navigation Satellite System (BDS) 北斗导航卫星系统[中]
Positioning, Navigation and Timing (PNT) 定位导航和授时
Coordinated Universal Time (UTC) 世界协调时，世界统一时间
medium Earth orbit (MEO) 中地球轨道
space segment 空间部分
control segment 控制部分
user segment 用户部分
GNSS constellation GNSS 星座
master control station 主控站
monitor station 监控站
GNSS receiver GNSS 接收机
navigation message 导航电文
satellite ranging 卫星测距
pseudorange 伪距

atomic clock 原子钟
clock error 钟差
tropospheric delay 对流层延迟
ionospheric delay 电离层延迟
multipath effect 多路径效应
ambiguity resolution 模糊度解算
Selective Availability (SA) 选择可用性
broadcast ephemeris 广播星历
precise ephemeris 精密星历
right ascension 赤经
ascending node 升交点
Coarse/Acquisition Code (C/A code) C/A 码
precise code 精码
Earth Centered Earth Fixed (ECEF) coordinate system 地心地固坐标系
International GNSS Service (IGS) 国际 GNSS 服务
absolute positioning 绝对定位
single-point positioning 单点定位
relative positioning 相对定位
static positioning 静态定位
static point positioning 静态单点定位
static relative positioning 静态相对定位
kinematic positioning 动态定位
kinematic single point positioning 动态单点定位
kinematic relative positioning 动态相对定位
real-time positioning 实时定位
Real-Time Kinematic (RTK) positioning 实时动态定位
carrier phase smoothed pseudorange method 载波相位平滑伪距方法
Precise Point Positioning (PPP) 精密单点定位
Standard Positioning Service (SPS) 标准定位服务
Precise Positioning Service (PPS) 精密定位服务
base station 基站
reference station 基站
roving station 流动站
reference receiver 基准接收机
roving receiver 流动接收机
receiver antenna 接收机天线
single reference station 单基站
network of multiple reference stations 多基站网络
network RTK 网络 RTK 技术
Continuously Operating Reference Station (CORS) 连续运行参考站系统

differential GNSS (DGNSS) 差分 GNSS
differential correction 差分改正
carrier phase correction 载波相位改正
real-time differential correction 实时差分改正
post-processed differential correction 后处理差分改正
Inertial Measurement Unit (IMU) 惯性测量单元

Unit 22 GIS Basics

Definition and Components

A geographic information system (GIS) is a computer system for capturing, storing, querying, analyzing, and displaying geospatial data. (Chang, 2018). The most powerful function of a GIS is to process, analyze and display geospatial data in different styles for different practical application which distinguishes GIS from other information systems.

Components of a GIS include: (1) Geospatial data: Geospatial data describe both the locations and characteristics of spatial features on the Earth's surface[①]. Spatial features such as buildings, roads, mountains, rivers, railroads, and political boundaries can be located (where they are) using either a geographic or a projected coordinate system together with descriptive characteristics (e.g., name, length, area or speed). (2) Hardware: GIS hardware mainly include input/output and process devices. Keyboards, digitizers, scanners, total stations, GNSS and mobile data collectors are examples of input devices. Output devices include displays, printers and plotters. Process devices include computer configuration/networks and data storages. (3) Software: GIS software include system software such as an operating system for Windows, Mac, or Linux, database software for data management and specialized GIS software such as ArcGIS, MAPGIS, SuperMap, etc. GIS software are the cores for execution of data input/edit, data management, data analysis, data display, and other tasks. (4) Professionals: GIS professionals include operational staff, technical professional staff, and management personnel. The professional knowledge and abilities are the keys for performance and quality assurance of the GIS project implementation. (5) Organization: GIS organization is responsible for GIS operation environment to meet integration of decision-making, the role and value, data collection and data standards.

Geospatial Data and Data Model

As mentioned above, geospatial data describe both locations and characteristics of spatial features on the Earth's surface. The former is called spatial (geospatial) data and later is called attribute data. Spatial data focus on the locations of the spatial features with georeference frame. Spatial data should be displayed in their proper places in relation to other spatial features in a GIS. Therefore georeferenced spatial data or geospatial data are also named for spatial data. For example, we use the longitude and latitude of a city to describe where the city is. Attribute data describe the characteristics of the spatial features and are identified and stored with the information about the spatial feature. For example, we use the city name, the population, the area and the ZIP code to describe the characteristics of the city or what the city is.

Spatial data or spatial features in the real world can be represented conceptually in following four identifiable types: Points, Lines, Polygons (Areas), and Surfaces. **Points** represent those discrete spatial features with no widths or lengths. **Lines** are conceptualized as occupying only a single dimension in space. Polygons have two dimensions with length and width. The terrain surfaces with undulating have three dimensions with length, width and height. Depending on the solution of the data and practical application, physical size in relation to the scale determines whether an object is represented by an area or by a point[②].

To represents spatial data in a GIS, either vector data model or raster data model can be used. The vector data model uses discrete objects defined by their x, y coordinates to represent spatial features on the Earth's surface. For example, the vector data model uses a given point defined by x and y locations in space to represent a point feature, a series of single coordinates connected to each other in space to represent a line feature which is one-dimensional and has the property of length besides location, a series of single coordinates that begin and end at the same location in space to represent a polygon feature which is two-dimensional and has the properties of size and perimeter besides location. While the raster data model uses a grid and grid cells to represent spatial features. For example, single discrete cells represent point features. Sequences of neighboring cells represent line features and collections of continuous cells represent polygon features. Raster models are useful for storing data that varies continuously. This continuous surface is divided into rows, columns, and cells. Cells are also called pixels and cell value corresponds to the attribute of the spatial feature at the cell location. For instance, each pixel value in an elevation map represents a specific height. Digital elevation data, satellite images, digital orthophotos, scanned maps, and graphic files are examples of raster data.

How Does GIS Work?

A GIS works by providing a powerful problem-solving tool to capture and edit data, store and manage the data, display, query and analyze the data. Let's take a closer look at how a GIS works.

Data Acquisition: Data acquisition is usually the first step in conducting a GIS project. Field surveys by total station or GNSS, aerial photogrammetry/remote sensing and scanning/digitizing existing maps are three common ways to collect spatial data. No matter vector or raster data, they have the spatial reference information so that they can align spatially with other data sets in a GIS.

Data Management: Data management in a GIS mainly focus on the vector data. There are two types of data — spatial data and attribute data need to be stored and managed. Although the georelational data model and the object-based data model handle the storage of spatial data differently, both operate in the same relational database environment. Therefore, data especially the attribute data management should be carefully designed and organized based on the relational database management system (RDBMS).

Data Display: Data display in GIS focus on the visual techniques to show the GIS production. Once all the geospatial data have been well organized and stored into a GIS system, the most common and effective application is to combine different data layers to produce a wide variety of individual maps. Mapmaking is a visually pleasing art to display or communicating geospatial data with map elements, cartographic symbology, color use, data classification, and even map animation.

Data Exploration: Data exploration is an important part of data analysis and data visualization in GIS. Data exploration lets people view and query geospatial data interactively and dynamically which involves spatial data query and attribute data query on maps and map elements. General statistical and graphic descriptive patterns in a data set and their possible relationships can also be viewed easily.

Data Analysis: Data Analysis is commonly thought of as the heart of the GIS. Buffering and overlay are two basic analysis tools for vector data. Local, neighborhood and zonal operations as well as physical distance measure operations are four basic tools for raster data[3]. Terrain analysis, viewshed and watershed analysis, spatial interpolation, geocoding, least-cost path analysis and networks analysis as well as GIS modeling are most powerful analysis tools in GIS[4].

Words and Expressions

geospatial [dʒiːəʊˈspeɪʃəl] adj.地理空间的
distinguish [dɪsˈtɪŋgwɪʃ] v.区别；辨别
attribute [əˈtrɪbjuːt] n.属性；品质；特征；加于；归结于
digitizer [ˈdɪdʒɪtaɪzə] n.数字转换器
plotter [ˈplɒtər] n.绘图仪
conceptional [kənˈsepʃənəl] adj.概念上的
undulating [ˈʌndʌleɪtɪŋ] adj.波状的；起伏的
raster [ˈræstər] n.[物]光栅；栅格
vector [ˈvektər] n.[数]向量；矢量；带菌者
pixel [ˈpɪksl] n.像素
orthophoto [ɔːrθəˈfəʊtəʊ] n.[摄]正射影像；正射像片

align [əˈlaɪn] v.使排列成行；排列
cartographic [ˌkɑːtəʊˈɡræfɪk] adj.制图的；地图的
symbology [sɪmˈbɒlədʒi] n.符号学；象征手法
buffering [ˈbʌfərɪŋ] n.缓冲（作用）；减震；阻尼；隔离
overlay [ˈəʊvərleɪ] n.覆盖；覆盖图
viewshed [ˈvjuːʃed] n.可视域
watershed [ˈwɔːtərʃed] n.分水岭；流域
interpolation [ɪnˌtɜːrpəˈleɪʃn] n.插补；内插法；插值
geocoding [dʒiːəʊˈkəʊdɪŋ] n.地理编码

☞ **Notes：**

①地理空间数据：地理空间数据描述了地球表面空间要素的位置和特征。
②多边形具有长度和宽度两个维度。绵延起伏的地表面具有长度、宽度和高度三个维度。根据数据的解决方案和实际应用，地物的实际尺寸及所需比例尺决定了运用面还是用点来表示。
③缓冲区分析和叠置分析是矢量数据的两个基本分析工具。局域运算、邻域运算、分区运算及自然距离量测运算是栅格数据的四大基本分析工具。
④地形分析、视域及流域分析、空间插值、地理编码、最小耗费路径分析和网络分析以及 GIS 建模是 GIS 最强大的分析工具。

Terms Highlights

geographic information system（GIS）地理信息系统
geospatial data 地理空间数据
spatial data 空间数据
attribute data 属性数据
vector data model 矢量数据模型
raster data model 栅格数据模型
point feature 点特征
line feature 线特征
polygon（area）feature 面特征
data acquisition（capture）数据获取（采集）
data management 数据管理
data display 数据显示
data exploration 数据探查
data analysis 数据分析
data classification 数据分类
data compression 数据压缩
georelational data model 地理关系数据模型

object-based data model 基于对象的数据模型
database management system (DBMS) 数据库管理系统
relational database management system (RDBMS) 关系数据库管理系统
buffering analysis 缓冲区分析
overlay analysis 叠置分析
local operation 局域运算
neighborhood operation 邻域运算
zonal operation 分区运算
physical distance measure operation 自然距离量测运算
terrain analysis 地形分析
viewshed analysis 视域分析
watershed analysis 流域分析
spatial interpolation 空间插值
least-cost path analysis 最小耗费路径分析
network analysis 网络分析
GIS modeling GIS 建模
digital terrain model (DTM) 数字地面模型
digital elevation model (DEM) 数字高程模型
digital surface model (DSM) 数字表面模型
regular grid 规则格网
triangular irregular network (TIN) 不规则三角网
topological relationship 拓扑关系
surface-fitting algorithm 曲面拟合算法
digital terrain visualization 数字地形可视化
chain code 链码
run-length encoding 游程编码

Unit 23 Digital Terrain Modeling

The creation of digital models of the terrain is a relatively recent development, and the introduction of the term Digital Terrain Model (DTM) is generally accredited to two American engineers, Miller C. and LaFlamme R. A., working at the Massachusetts Institute of Technology during the late 1950s (Petrie and Kennie, 1990). The definition given by them is as follows:

"A statistical representation of the continuous surface of the ground by a large number of selected points with known X, Y and Z coordinates in an arbitrary coordinate field."

Digital Terrain Modeling is the electronic process of representing topography in three dimensions. It consists of a number of surface points that are representative of the terrain and designate the positions of points in relation to a common reference frame. In common usage, the X and Y coordinate fix the horizontal position of the point, and Z refers to the elevation. A DTM may involve a range of geographical elements and natural features such as rivers or ridge lines and may include derived data about the terrain such as slope, aspect, visibility, etc.

The term Digital Elevation Model (DEM) specifically relates to the height above a datum and the absolute elevation of the points contained in the model. In any case, the term usually refers to the creation of a regular array of elevations, normally in a square grid, over the terrain. The manipulation of the data in such a form in a computer system is straightforward since a DEM is essentially a two-dimensional matrix.

DTM is a multistep process that is made up of the following sequence of tasks [C.P. Lo and Albert K.W.Yeung, 2002]:

(1) Digital terrain data sampling is the structuring and acquisitions of digital terrain data by photogrammetric, cartographic, and field survey methods.

(2) Digital terrain data processing is the manipulation of digital terrain to ensure their usability by GIS.

(3) Digital terrain data analysis involves the use of algorithms and procedures that restructure digital terrain data into useful geographic information.

(4) Digital terrain data visualization entails the development of algorithms and methods that allow the effective display of the terrain to assist in spatial problem solving and decision making.

(5) Digital terrain data applications comprise the practical use of DTM in different fields of science and technology.

Digital Terrain Data Sampling

There are two approaches to digital terrain data sampling: systematic and adaptive. In systematic terrain data sampling, elevation points are measured at regularly spaced intervals (Regular Grid). The result is a matrix of elevation values that is usually referred to as a digital elevation model (DEM). The locations of elevation points in DEM are implicit in the data model. When adaptive sampling method is used, elevation measurements are made at selected points that are assumed to be representative of the terrain. The result is a collection of irregularly distributed elevation values that must be properly structured before they can be used for further processing. Since the method of triangulation is used to build the spatial framework for storing the elevation values, the data collected by this approach are referred to as a triangular irregular network (TIN). TIN is a model of the vertices of triangles generating from the distributed data set, which form uniquely spaced nodes.① Unlike the grid (regularly spaced intervals), the TIN provides for dense information in complex areas and sparse information in homogeneous areas. The TIN data sets provide topographical value among points and their neighboring triangles. The TIN data model is distinct from the DEM data model in two important ways: each and every sample point in a TIN has an (X, Y) coordinate and an elevation, or Z value; the TIN data model may include explicit topological relationship between points and their proximal triangles.

Digital terrain data may be acquired by a variety of methods, depending on factors such as the location and size of the area of interest, the purpose of the terrain modeling, and the technical resources available.② Generally speaking, ground survey methods are most suitable for large-scale terrain modeling for engineering and mining applications. At smaller scales covering larger geographic areas, photogrammetric methods are always used. Currently terrain information may be acquired by remote sensors on airborne platforms. However, as a vast amount of terrain data are already in existence in topographic maps, many national mapping agencies tend to acquire digital terrain data by digitizing existing maps. Digital terrain data sets obtained in this way are usually of a small scale and have a national or regional coverage.

Data Processing and Analysis

The core activities of digital terrain modeling involve three typical phases: processing terrain data to ensure that they are optimized for storage and application; performing analysis to convert topographic attributes (elevation, slope, aspect, profiles curvature, and catchments (area) derived from DEMs or TINs into useful terrain information; and presenting the terrain information to the user in an easily understandable manner. Mesh simplification is the process by which a TIN model is constructed from DEM data. The objective is to extract from a DEM the topographically important elevation points to form a TIN with the minimum number of points possible, while at the same time preserving the maximum amount of information about terrain structure.[3] This is an essential function in digital terrain data processing because it allows the user to take advantage of both the DEM and TIN models. The DEM approach is more suitable than the TIN approach for automatic digital terrain data sampling. With mesh simplification techniques and powerful computers, it is now possible to acquire digital terrain data with DEM approach. The data are then processed to form TINs to optimize storage and modeling efficiency.

Interpolation is the process by which elevation values of one or more points in geographic space are used to produce estimated values for positions where elevation information is required. It is used for contouring and for the generation of DEMs from selectively or randomly sampled elevation points. A surface-fitting algorithm is commonly used to improve the result of terrain modeling. It may be linear or nonlinear, depending on the order of the polynomial equations used for estimating the elevation of the required point.

Digital Terrain Visualization

The ultimate aim of DTM is to present relevant terrain information about a given geographic space that results from the analysis of the characteristics of its topography and related spatial phenomena.[4] Visualization is therefore an integral component of DTM, forms the perspective of both process and technology.

There are numerous well-developed techniques for digital terrain visualization. According to the dimension of the graphical display, digital terrain visualization can be classified as: two-dimensional, two-and-a-half-dimensional, three-dimensional, and multidimensional. Contour lines are the most conventional, and probably still the most commonly used method of digital terrain visualization. Using contour line is a quantitative way of representing three-dimensional terrain in two dimensions because numerical measurements of elevation may be readily made on the display. A two-and-a-half-dimensional display is basically an isometric model. In such a model, the Z attribute associated with an X, Y location is projected onto an X, Y, Z coordinates reference system. This transforms the map of Z attributes for an X, Y position so that each Z attribute defines a position on the Z axis, creating a surface that is perceived as three-dimensional. A three-dimensional terrain model is a solid model in which many X, Y, Z data points are used to form a solid structure that may be visualized in a perspective view. Unlike a two-and-a-half-dimensional view, which presents only a pseudo perspective of the terrain, a three-dimensional terrain model is an analog for the physical space in nature as perceived by an observer. Three-dimensional terrain-modeling allows the full specification of three-dimensional operations on the objects and phenomena within the constraints of the geometrical model used. It represents one of the most exciting developments in GIS technology.

Application of DTM

Once a DTM has been created, contours, profiles, volumes between surfaces and three-dimensional displays are available. In the last several years, there has been a tremendous growth in the application of DTMs, not only in the traditional fields of geography, surveying and mapping, and earth and environmental sciences, but also in landscape design, biodiversity analysis, environmental impact analysis and site selection for telecommunication facilities.

Words and Expressions

cost-effective [ˌkɒstɪˈfektɪv] *adj.* 费用低廉的
spaceborne [ˈspeɪsbɔːn] *adj.* 宇宙飞行器上的；卫星（飞船）上的
accredited [əˈkredɪtɪd] *adj.* 可接受的；可信任的；公认的；质量合格的
Massachusetts [ˌmæsəˈtʃuːsɪts] *n.* 马萨诸塞州
representative [ˌreprɪˈzentətɪv] *n.* 代表 *adj.* 典型的；有代表性的
ridge line 屋脊线
multistep [ˈmʌltɪstep] *adj.* 多步的；多级的
usability [ˌjuːzəˈbɪləti] *n.* 可用性
framework [ˈfreɪmwɜːk] *n.* 构架；框架；结构
node [nəʊd] *n.* 节点
sparse [spɑːs] *adj.* 稀少的；稀疏的
topological [ˌtəʊpəˈlɒdʒɪkəl] *n.* 拓扑的
proximal [ˈprɒksɪməl] *adj.* 最接近的
catchment [ˈkætʃmənt] *n.* 集水；集水处；汇水
simplification [ˌsɪmplɪfɪˈkeɪʃn] *n.* 简化
mesh [meʃ] *n.* 格网；网孔；网丝；网眼；圈套；陷阱
polynomial [ˌpɒliˈnəʊmiəl] *adj.* [数] 多项式的
perspective [ˈpərspektɪv] *n.* 透视画法；透视图；远景；观察
isometric [ˌaɪsəˈmetrɪk] *adj.* 等大的；等容积的
tremendous [trəˈmendəs] *adj.* 极大的；巨大的
telecommunication [ˌtelɪkəˌmjuːnɪˈkeɪʃn] *n.* 电信；长途通信；无线电通信；电信学

☞ Notes：

① 不规则三角网是由分布数据集生成的三角形顶点的模型，这些顶点组成了特有的空间节点。

② 数字地面数据的获得有多种方法，这取决于多种因素，如感兴趣区域的位置和大小，地面建模的目的以及可获取的技术资源。
③ 目的在于从数字高程模型中提取重要的地形高程点，来形成一个具有最少点的不规则三角网，同时最大限度地保留地形结构的信息。
④ 数字地面模型的最终目的是为了展示给定地理空间的相关地面信息，分析它的地形特征和相关的空间现象。

Terms Highlights

geographic information system(GIS) 地理信息系统
image data 图像数据
vector data 矢量数据
raster data 栅格数据
data capture 数据采集
attribute data 属性数据
data classification 数据分类
data compression 数据压缩
data transmission 数据传输
spatial data transfer 空间数据转换
geographic information communication 地理信息传输
spatial analysis 空间分析
place-name database 地名数据库
spatial database management 空间数据库管理系统
spatial data infrastructure 空间数据基础设施
digital terrain model(DTM) 数字地面模型
digital elevation model (DEM) 数字高程模型
digital surface model(DSM) 数字表面模型
triangular irregular network (TIN) 不规则三角网
topological relationship 拓扑关系

Unit 24　Photogrammetry Overview

Development of Photogrammetry

Photogrammetry may be defined as the art, science, and technology of obtaining reliable information about physical objects and the environment, through processes of recording, measuring, and interpreting images and patterns of electromagnetic radiant energy from noncontact sensor systems (American Society for Photogrammetry and Remote Sensing, 2015). It has a history as long as photography itself. The French officer Aimé Laussedat was the first to use photography for surveying in 1851, and the Prussian architect Albrecht Meydenbauer coined the term *photogrammetry* in 1867.

The main concern of photogrammetry is to make precise measurements of three-dimensional objects and terrain features from two-dimensional images. Two-dimensional images, recorded by cameras, are the perspective projection of three-dimensional object space to the image plane. It can hardly make precise measurements from only one two-dimensional image. Usually, corresponding image points from two or more images are used to calculate precise three-dimensional position of objects based on forward-intersection theory. During the last 170 years, the theory of photogrammetry in topographic mapping has been well-established.

The development of photogrammetry has passed through the phases of analog photogrammetry and analytical photogrammetry, and has now entered the phase of digital photogrammetry. In the phase of analog photogrammetry, purely optical or mechanical instruments were used to reconstruct three-dimensional geometry by simulating the intersection of homologous rays from two cameras during exposure. The main products during this phase were topographic maps. With the development of computing technology, it moved to computer-aided analytical photogrammetry phase in the 1950s. In this phase, rigorous mathematical calculation was performed to solve the relationship between coordinates of image points measured in image plane coordinate system and coordinates of object points in ground coordinate system. Outputs of the analytical photogrammetry can be topographic maps, but can also be digital products such as digital line graphs (DLGs) and digital elevation models (DEMs). The modern photogrammetry based on computer vision is called digital photogrammetry, which deals with digitized images scanned from photographs and digital imagery directly captured by digital cameras. The primary results are in digital forms such as DEMs and digital orthophoto maps (DOMs); therefore, digital photogrammetry is sometimes called full digital (softcopy) photogrammetry.

Photogrammetry Classifications

A photogrammetry system is composed of hardware and software. The hardware components are responsible for data acquisition. Cameras are the core and basic sensors in photogrammetry hardware used to take images of targets. Besides cameras, position and orientation system (POS) is usually used to obtain position and attitude data to facilitate data post-processing. According to where the cameras are mounted, photogrammetry can be divided into aerial photogrammetry and terrestrial photogrammetry.

Aerial photogrammetry takes photographs from airborne platforms, that is, the camera can be placed on a fixed-wing aircraft, helicopter, and unmanned aerial vehicle (UAV). The workflow of aerial photogrammetry includes flightline planning, data acquisition, control surveying and image processing. Flightline planning creates flightlines with a specific scale and accuracy based on the camera parameters and the specifications. Data acquisition captures images according to the planned flightline with the designed platform and cameras. The aerial photography is linked with the ground by control points. The control surveying is vitally important since it determines the accuracy of the final products. Image processing mainly deals with tie points matching, automatic aerial triangulation (AAT), dense image matching and ortho-rectification. The final products of aerial photogrammetry includes DEM, DOM and DLG.①

Aerial photographs can be generally classified as being vertical or oblique. Traditionally, the aerial photographs are taken with the camera axis directed toward the ground as vertically as possible. When viewed in stereo, vertical photographs can provide height information of ground features. This kind of vertical photogrammetry is often used in topographic mapping, land-use planning, and aerial archaeology. Oblique photogrammetry technology is an emerging technology capturing images from the vertical direction and four inclination directions with multiple sensors. In this way, top and side views of buildings with high-resolution textures can be synchronously acquired, facilitating 3D real scene construction for smart city.② Oblique photogrammetry can be used for urban management, emergency command, power line inspection, and real estate taxation.

Terrestrial photogrammetry is also known as *close-range photogrammetry*, in which the camera is handheld or mounted on a tripod with an object-to-camera distance of less than 100m. Similar to aerial photogrammetry, the workflow of terrestrial photogrammetry includes project design, image acquisition, control surveying, and image processing. In contrast to aerial photogrammetry, control surveying is not essential for terrestrial photogrammetry. The purpose of this method is usually not to create topographic maps, but to get physical and geometric information of ground objects for architectural, industrial, and medical use.

Data Processing Workflow

Professional photogrammetry software has been developed for image processing and results output, in which tie points matching, bundle adjustment, dense image matching and ortho-rectification are essential operations.

The aim of tie points matching is to find correspondences in overlapping areas of image pairs. In the early days, the tie points are identified by the manual intervention. Nowadays, repeatable and stable interest points are first extracted from each image, then the neighborhood of each interest point is encoded with a descriptor, and tie points between image pairs can be automatically matched by comparing descriptor values of the interest points.[3] Scale invariant feature transform (SIFT) is perhaps the most famous and widely used tie points matching algorithm.

Bundle adjustment is the mainstream method of AAT used to find the optimal pose of each image and ground coordinates of pass points. Bundle refers to the bundle of light rays leaving each three-dimensional feature and converging on camera center. The correspondences obtained in tie points matching serve as the input of bundle adjustment. With initial values, bundle adjustment simultaneously refines the exterior orientation parameters, the three-dimensional ground coordinates, and even the interior parameters of cameras and distortion coefficients of lens according to an optimality criterion involving the corresponding image projections of all three-dimensional points.[4]

One can only get sparse reconstruction of the scene from bundle adjustment. To get dense reconstruction, dense image matching is performed to find the right correspondence for every pixel in an image. To achieve this goal, local methods, global methods, or semi-global methods can be used. Local methods compute the disparity at a given point based on the intensity values within a finite region, and smoothing assumptions are implied. These approaches can be very efficient but sensitive to sudden depth variations and occlusions. On the other hand, global methods make explicit smoothness assumptions and then solve for a global optimization problem using an energy minimization approach. Global methods can achieve best performance, but it is time-consuming. Semi-global methods perform line optimization along multiple directions, and they are widely used due to predictable runtime and acceptable quality.

Orthophoto is an important product of photogrammetry obtained by ortho-rectification. The raw images captured by photogrammetry system suffer from the effects of perspective (tilt displacement) and relief displacement due to the pinhole camera model and topographic relief. Ortho-rectification is to remove the perspective and relief effects from the raw images to create a planimetrically correct image. The resultant orthorectified image has a constant scale wherein features are represented in their "true" positions. Distances, angles, and areas can be directly measured from orthophoto.

Applications

Photogrammetry can be used for topographic survey, providing basic geographic data for urban planning, agriculture, forestry, mining, military, and other departments. It has proven to be an accurate and cost-effective method. The accuracy depends on distances, pixel density of the images, lighting condition, and the algorithm of the processing software. Photogrammetry is also widely used for non-topographic survey such as industry, engineering, and medicine to obtain precise three-dimensional data. In industrial manufacturing, geometric measurements of mechanical components based on photogrammetry are crucial to quality control and assembly. Moreover, photogrammetry can be utilized to assess body in white (BIW) structural movement under force in automotive testing. With photogrammetry, oil and gas companies can get an accurate aerial view of the pipelines for gas and petrol as well as their corridors. In recent years, photogrammetry has extended from aerial to terrestrial, and now it is extended underwater. Although underwater photogrammetry is still in an exploratory phase, underwater terrain and models are needed in the fields of environmental monitoring, archaeology, infrastructure inspection, and so on.

Words and Expressions

photogrammetry [fəutəʊˈɡræmɪtri] n.[测]摄影测量学；摄影测量法
noncontact [ˈnɒnkənˈtækt] n.无接点 adj.非接触(式)的；无接触的
photography [fəʊtəˈɡrɑːfi] n.摄影；摄影术；照相术
coin [kɔɪn] n.硬币；金属货币 v.创造(新词语)；首次使用；铸币；制造
corresponding [ˌkɒrɪˈspɒndɪŋ] adj.对应的；相关的；符合的；一致的
exposure [ɪkˈspəʊʒə(r)] n.曝光；暴露；揭露；面临
bundle [ˈbʌndəl] n.束；捆；包；一套；一批
ortho-rectification [ˌɔːθəʊ-ˌrektɪfɪˈkeɪʃn] n.正射纠正；正射校正
descriptor [dɪˈskrɪptə] n.[计]描述符；描述子
converge [kənˈvɜːdʒ] v.汇聚；集中；(向某一点)相交；[数]收敛
exterior [ɪkˈstɪəriər] adj.外部的；外面的；外表的；户外的 n.外观；外表
interior [ɪnˈtɪəriər] adj.内部的；里面的；国内的；内政的 n.内部；里面；内陆
disparity [dɪˈspærəti] n.明显差异；悬殊；视差
occlusion [əˈkluːʒn] n.闭塞；阻塞；遮挡
BIW (body in white) 白车身(完成焊接但未涂装之前的车身)

☞ **Notes**：

① 图像处理主要涉及连接点匹配、自动空中三角测量、影像密集匹配、正射纠正。航空摄影测量的最终产品包括数字高程模型(DEM)、数字正射影像(DOM)以及数字线划图(DLG)。

② 倾斜摄影测量技术是一种新兴技术，它利用多个传感器从1个垂直方向和4个倾斜方向获取影像。通过这种方式，可以同步获取建筑物顶面及侧视的高分辨率纹理，为智慧城市的实景三维构建提供了便利。

③ 现如今，我们首先从每幅影像中提取可重复且稳定的兴趣点，然后将每个兴趣点的邻域用描述符编码，再通过比较兴趣点的描述符值就可以自动匹配影像对之间的连接点。

④ 给定初值后，光束法平差根据所有三维点同名光线最佳交会的最优准则，对外方位参数、三维地面坐标，甚至是相机内参数和镜头畸变系数同时进行精化。

Terms Highlights

analog photogrammetry 模拟摄影测量
analytical photogrammetry 解析摄影测量
digital photogrammetry 数字摄影测量
full digital (softcopy) photogrammetry 全数字摄影测量；软拷贝摄影测量
aerial photogrammetry 航空摄影测量
aerial photography 航空摄影
vertical photograph 垂直像片
vertical photogrammetry 垂直摄影测量
oblique photogrammetry 倾斜摄影测量
photographic scale 摄影比例尺
flight altitude 航高
photographic principal distance 摄影主距
photo tilt 像片倾斜
stereopair, stereo photopair 立体像对
terrestrial photogrammetry 地面摄影测量
close-range photogrammetry 近景摄影测量
stereophotogrammetry 立体摄影测量
homologous points, corresponding image points 同名像点
homologous rays 同名光线；同名射线
internal orientation 内定向
relative orientation 相对定向
absolute orientation 绝对定向
interior parameters 内参数
distortion coefficients 畸变系数

exterior orientation parameters 外方位参数
tie points matching 连接点匹配
scale invariant feature transform (SIFT) 尺度不变特征变换
pass point 加密点
analytical aerotriangulation 解析空中三角测量；电算加密
automatic aerial triangulation (AAT) 自动空中三角测量
sparse reconstruction 稀疏重建
dense image matching 影像密集匹配
ortho-rectification 正射纠正
block adjustment 区域网平差
bundle adjustment 光束法平差
image mosaicing 影像镶嵌
image rectification 影像纠正
digital line graph (DLG) 数字线划图
digital orthophoto map (DOM) 数字正摄影像图
picture element/ pixel 像元/像素
intensity value 强度值；亮度值
CCD camera CCD 相机
position and orientation system (POS) 定位定向系统
unmanned aerial vehicle (UAV) 无人机；无人飞行器
analytical plotter 解析测图仪
stereocomparator 立体坐标量测仪
reflecting stereoscope 反光立体镜
stereo glasses 立体镜

Unit 25 Fundamentals of Remote Sensing

As defined by the U.S. Geological Survey (USGS, 2017, https://www.usgs.gov/faqs/what-remote-sensing-and-what-it-used).

Remote Sensing (RS) is the process of detecting and monitoring the physical characteristics of an area by measuring its reflected and emitted radiation at a distance (typically from satellite or aircraft).

Even though the term "remote sensing" was first used to describe the field in the 1960s, the history of remote sensing begins with photography. In 1839, Louis Daguerre reported his discovery of practical photography technology. Since then, various methods such as mounting a camera on balloons and pigeons, have been used to take photographs from the air. The invention of airplane by Wright brothers (Orville Wright and Wilbur Wright) made it possible to take photographs from an airplane. The first aerial photographs were taken from an airplane in 1909 by Wilbur Wright. During the First World War and the Second World War, large numbers of photographs were taken from aircraft for military reconnaissance purposes. At that time, remote sensing data interpretation technology was developed to detect objects such as army installations. With the advent of satellite "Sputnik" in 1957, the possibility of putting remote sensors on orbiting spacecraft was realized. The American Landsat 1, originally named Earth resources technology satellite-1 (ERTS-1), launched on July 23, 1972, is one of the most famous satellite remote sensing platforms. Other famous satellite remote sensing platforms include French SPOT series, commercial WorldView series, et al. Remote sensing can be used in various fields, such as resource census, water quality monitoring, land use survey, natural disaster monitoring, weather forecast, agricultural fine monitoring and so on.

Fundamental Types of Remote Sensing

Remote sensors are the core of a remote sensing system. According to whether the remote sensors actively emit radiation or not, remote sensing can be categorized into *passive remote sensing* and *active remote sensing* (Figure 1). For passive remote sensing, the remote sensors do not actively emit radiation itself and use the solar radiation reflected by the Earth surface or the thermal infrared (IR) radiation emitted by the Earth. On the contrary, active remote sensing sensors emit microwave or laser pulses actively and then collect the radiation reflected by the Earth's surface. Typical active remote sensing includes radio detection and ranging (Radar) and light detection and ranging (LiDAR) scanning.

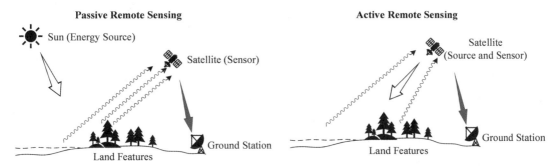

Figure 1

Remote sensing can also be divided into *terrestrial remote sensing*, *airborne remote sensing*, and *satellite remote sensing* based on where the remote sensors are placed. For terrestrial remote sensing, the sensors are mounted on a tripod, a car, or a monitoring tower, et al. For airborne remote sensing and satellite remote sensing, the sensors are mounted on a UAV or aircraft, and satellite, respectively. In terrestrial remote sensing and airborne remote sensing, cameras are the most widely used sensors. Since cameras record the images of the objects at an instance of exposure, such remote sensing is called *photographic remote sensing*. For most modern satellite remote sensors, they obtain the images of objects pixel by pixel or line by line, such remote sensing is called *non-photographic remote sensing*.

Remote sensing uses the radiation reflected or emitted by the target area to detect and monitor the physical characteristics of the target area. According to the electromagnetic spectrum used (Figure 2), remote sensing can be categorized into *ultraviolet remote sensing*, *visible light remote sensing*, *infrared remote sensing*, and *microwave remote sensing*. Infrared remote sensing can be further divided into *near infrared (NIR) remote sensing* and *thermal remote sensing*. The gamma ray and x-ray are seldom used in remote sensing.

Remote Sensing Sensors

Remote sensing sensors are the core components of any remote sensing systems, which can be categorized into passive sensors (when the reflection of sunlight is detected by the sensor) and active sensors (when a signal is emitted to the object and its reflection detected by the sensor).

Passive sensors: The oldest remote sensing technology is based on cameras, providing a still image in the visible range. Films were originally used, but they have been totally replaced by digital sensors in the past decades, among which large-format cameras are the primary source for optical imagery in airborne surveying. Imagery acquired by panchromatic sensors provides good spatial resolution for geometric feature extraction, but generally lacks the spectral information needed for classification. The professional systems are typically based on individual camera heads with high quality filters to produce highly accurate RGB, NIR, etc., images that can be merged with panchromatic imagery to obtain high resolution multispectral (MS) images. For better classification performance, higher multispectral resolution is needed, for instance, 8~15 bands are frequently sensed, which is the typical case for MS sensors on satellites. For the most demanding applications, for example, where material signatures must be identified, hundreds of spectral bands are usually used and these data are called hyperspectral images (HSIs), also referred to as image cube.①

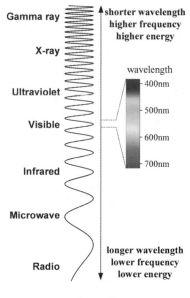

Figure 2

Active sensors: Radar and LiDAR are two typical types of active sensors. Radar is a detection system using radio waves to determine the distance, angle, and radial velocity of objects relative to the site. Radar is a century old concept, though radar imaging started to develop decades later. Radar has good resistance to weather conditions, and its cloud penetration capability should be mentioned. The first side-looking airborne radar systems (SLARs) were introduced around 1950, shortly followed by synthetic aperture radar (SAR). LiDAR determines the range by targeting an object with a laser and measuring the time for the reflected light to return to the receiver. LiDAR is now widely available from space down to terrestrial systems. Satellite LiDAR systems are predominantly used for atmospheric measurement, such as the aerosol content of the atmosphere. In contrast, airborne and ground LiDAR systems are primarily used for mapping surfaces.

Remote Sensing Platforms

Platforms refer to the structures or vehicles on which remote sensing instruments are mounted. There are typically three types of remote sensing platforms: terrestrial platforms, airborne platforms, and satellite platforms. The platforms determine many attributes, including the distance from object to sensor, periodicity of image acquisition, and location and extent of coverage.

Terrestrial platforms: The ground-based remote sensing system is often used to measure the quantity and quality of light coming from the sun or for close range characterization of objects. The terrestrial platforms can be static or mobile platforms, such as handheld devices, tripods, towers, cranes, and vehicles. Permanent platforms are typically used for long-term monitoring of terrestrial features and atmospheric phenomenon. Towers can be built on site to provide a reasonably stable platform, so a range of measurements can be taken from the forest floor, through the canopy and from above the canopy.

The basic concept of mobile mapping systems (MMS) is to use sensors installed on a moving vehicle, such as cars, trucks, and rail cars, to acquire geospatial data in a highly effective way over transportation corridors at normal travel speed. <u>After the slow start, the technological advancements have quickly propelled mobile mapping, and the term mobile mapping technology (MMT) was introduced about a decade ago after the Fourth International Symposium on Mobile Mapping Technology to reflect the significant performance improvements.</u>② These systems are usually equipped with multiple digital cameras and laser sensors as well as application specific sensors. Lots of MMT systems are currently used by the large Internet map data providers, for example, Google runs an estimated fleet of 1000+ vehicles.

Airborne platforms: Before commercial satellite remote sensing data is accessible, airborne remote sensing used to be the primary source of geospatial data. The first aerial images were acquired from a balloon, but balloons are rarely used today because they are not very stable, and the course of flight is not always predictable. At present, airplanes are the most common airborne platform, including fixed-wing aircrafts, helicopters, and drones. Helicopters are usually used for low altitude applications where the ability to hover is required. Mid-altitude and high-altitude aircrafts can obtain greater areal coverage more quickly than low altitude platforms. In general, the higher an aircraft can fly, the more stable a platform it is, but correspondingly more costly to operate and maintain. Since government restrictions on satellite imagery resolution that images with better than 30cm resolution cannot be commercially sold, airborne remote sensing is still an important complement to satellite remote sensing.

A typical remote sensing unmanned aircraft systems (UAS) is composed of gas or electric propelled aircraft, navigation system, imaging sensors, and communication system. Both fixed and rotating wing solutions are used in UAS practice, and combined solutions are now adopted to take the full advantages of flying times and maneuverability. The changes that have happened in the airborne platforms are primarily related to the introduction of new sensors, such as oblique photogrammetry systems, LiDAR and interferometric synthetic aperture radar (InSAR).

Satellite platforms: The most stable spaceborne platform is satellite, and satellite remote sensing have been around for 50 years. Landsat-1 was launched in 1972, followed by SPOT-1 in 1986 and IKONOS in 1999. IKONOS heralded the era of commercial satellite systems, and commercial satellite remote sensing images are now accessible and price affordable. The payload for remote sensing satellites can include photographic systems, electro-optical sensors, microwave or LiDAR systems. These platforms can acquire large areas of data in a short time, which can be used to monitor earth resources, atmospheric dynamics, and other applications.

The satellite remote sensing systems in the early stage were single satellite-based systems, and these systems are generally transitioning from the single sensor model to the cooperative sensing approach. Satellites fly in tandem or even constellations allowing for shorter revisit times. Landsat, SPOT and GeoEye/WorldView families represented the first such constellations. <u>A very specific type of multi-satellite constellation is the "train" where satellites equipped with different sensors are following each other on the same orbital "track" with a short time separation. The most known is the *A-Train*, where A stands for Afternoon constellation, consisting of six satellites: Aqua (2002), Aura (2004), CALIPSO(2006), CloudSat(2006), GCOM-W1(2012), and OCO-2(2014).</u>③ As of January 2024, the Afternoon constellation includes only GCOM-WI and OCO-2. The most recent development is the launch of flocks of nano/microsatellites in the same orbit, allowing for frequent observations.

Words and Expressions

emit [ɪˈmɪt] *v.* 发出；散发(辐射、光、热、气等)；发行
radiation [ˌreɪdɪˈeɪʃn] *n.* 辐射的热(或能量)；放射线；辐射形进化；放射疗法
census [ˈsensəs] *n.* 人口普查；(官方的)统计
thermal [ˈθɜːml] *adj.* 热的；热量的；(衣服)保暖的 *n.* 上升的热气流
ultraviolet [ˌʌltrəˈvaɪəlɪt] *adj.* 紫外的；紫外线的；利用紫外线的 *n.* 紫外光；紫外辐射
visible [ˈvɪzəbl] *adj.* [物理][光]可见的；看得见的；明显的 *n.* 看得见的事物
panchromatic [ˌpænkrəʊˈmætɪk] *adj.* [摄影]全色的
spectral [ˈspektrəl] *adj.* [光]光谱的；幽灵似的；鬼怪的
merge [mɜːdʒ] *v.* (使)融合；(使)合并；融入
multispectral [ˌmʌltiˈspektrəl] *adj.* 多光谱的；多谱的
hyperspectral [ˌhaɪpərˈspektrəl] *adj.* 高光谱的
penetration [ˌpenɪˈtreɪʃn] *n.* 穿透；渗透；进入
synthetic [sɪnˈθetɪk] *adj.* 合成的；人造的；综合(型)的 *n.* 合成物；合成纤维(织物)
predominantly [prɪˈdɒmɪnəntli] *adv.* 绝大多数地；主要地；占主导地位地
aerosol [ˈeərəsɒl] *n.* [物化]气溶胶；气雾剂；雾化器；浮质
periodicity [ˌpɪərɪəˈdɪsɪti] *n.* [数]周期性；频率；定期出现
handheld [ˈhændheld] *adj.* 手持式的；掌上的
crane [kreɪn] *n.* 起重机；吊车；鹤 *v.* 吊运；伸长(脖子)看
canopy [ˈkænəpi] *n.* 树冠层；天篷；顶篷 *v.* 用顶篷遮盖
propel [prəˈpel] *v.* 推进；推动；驱使；驱动
predictable [prɪˈdɪktəbl] *adj.* 可预测的；可预见的；意料之中的

drone [drəʊn] n.(遥控的)无人驾驶飞机(或导弹);雄蜂;嗡嗡声 v.嗡嗡叫
maneuverability [ˌmenuːvərəˈbɪlɪti] n.[航][车辆][船]机动性;可操作性
interferometric [ˌɪntəfərəʊˈmetrɪk] adj.干涉仪的;用干涉仪测量的;干涉测量的
herald [ˈherəld] v.预示;是……的前兆;宣布 n.预兆;使者
payload [ˈpeɪləʊd] n.(飞机、船只的)载荷;有效负荷;装载量
tandem [ˈtændəm] n.串联;双人自行车;并行 adj.纵列的
flock [flɒk] n.群;一大群人 v.聚集;群集;蜂拥
microsatellite [ˈmaɪkrəʊsætəlaɪt] n.微卫星

☞ **Notes**:

①对于那些要求最苛刻的应用,例如必须识别材料特征,通常使用数百个光谱带,由此获得的这些数据称为高光谱图像(HSI),也称为图像立方体。
②在缓慢的起步之后,技术的进步迅速地推动了移动测量的发展。大约十年前,在第四届国际移动测量技术研讨会之后,引入了移动测量技术(MMT)一词来反映这些显著的性能提升。
③一种非常特殊的多卫星星座类型是"卫星编队":搭载不同传感器的卫星在同一轨道上以较短的时间间隔相互跟随。最著名的是 *A-Train*(下午卫星编队),其中 *A* 代表下午星座,曾经由6颗卫星组成:Aqua(2002),Aura(2004),CALIPSO(2006),CloudSat(2006),GCOM-W1(2012)和OCO-2(2014)。

Terms Highlights

remote sensor 遥测传感器;遥感器
passive remote sensing 被动式遥感;无源遥感
active remote sensing 主动式遥感;有源遥感
terrestrial remote sensing 地面遥感
airborne remote sensing 航空遥感;机载遥感
satellite remote sensing 卫星遥感
photographic remote sensing 摄影方式遥感
non-photographic remote sensing 非摄影方式遥感
electromagnetic radiation 电磁辐射
electromagnetic spectrum 电磁波频谱;电磁波谱;电磁光谱
ultraviolet remote sensing 紫外遥感
visible light remote sensing 可见光遥感
infrared remote sensing 红外遥感
thermal infrared (TIR) 热红外
near infrared (NIR) 近红外
microwave remote sensing 微波遥感
hyperspectral image (HSI) 高光谱图像

image cube 图像立方体
radio detection and ranging (Radar) 雷达
side-looking airborne radar system (SLAR) 机载侧视雷达系统
synthetic aperture radar (SAR) 合成孔径雷达
interferometric synthetic aperture radar (InSAR) 干涉合成孔径雷达
light detection and ranging (LiDAR) 激光雷达
Earth resources technology satellite-1 (ERTS-1) 地球资源技术卫星 1 号
unmanned aircraft system (UAS) 无人机系统
airborne surveying 航空测量
mobile mapping system (MMS) 移动测量系统
mobile mapping technology (MMT) 移动测量技术

Unit 26　UAV Mapping Technology

An unmanned aerial vehicle (UAV) or a drone is a reusable aircraft that does not carry a human operator, uses aerodynamic forces to provide lift, and can fly autonomously or be remote-controlled. UAV mapping technology refers to the technical means of data acquisition through various sensors onboard, such as high-resolution digital cameras and laser scanners, data processing using corresponding software, and results output with a certain precision.① Comparatively speaking, traditional mapping is a highly time-consuming effort, while UAV mapping can significantly reduce the turnaround time for gathering data and representing it with a map.

The first pilotless aircraft was developed for the British Royal Flying Corps as target drone in 1917, naming aerial target (AT). Reconnaissance UAVs began to be used in the 1960s and were first deployed on a large scale in the Vietnam War. By the end of the 20th century, UAVs were still mainly for military use, and gradually expanded to civilian applications. The developments of civilian UAVs have stepped into the industrialization stage since 2010s, such as in the fields of aerial photography, surveying and mapping, express transportation, news reporting, and disaster relief. In UAV mapping, hundreds of aerial images are acquired and then stitched together digitally with specialized mapping software to make a larger more accurate composite image. It is a powerful supplement to traditional aerial photogrammetry, which has obvious advantages in the rapid acquisition of high-resolution images in small areas and mountainous areas.

Unmanned Aircraft System (UAS)

The term UAS was adopted by the US Department of Defense (DoD) and the Federal Aviation Administration (FAA) according to their UAS roadmap 2005-2030. This system includes elements such as flying platform, flight control system, ground control system, sensor system, data links and other support equipment.

Flying Platform: The UAVs can be fixed-wing, rotary-wing or multirotor, and hybrid. Fixed-wing UAVs have flat wings and use less energy than rotary-wing UAVs, which are better adapted to long distances.② The lowest end fixed-wing UAVs have a flight endurance of only a few dozen minutes, while the most expensive models can last up to 24 hours. Fixed-wing UAVs can also be faster, fly at higher altitudes, and carry heavier loads, so they can cover larger areas in the same amount of time and collect as much information as possible. As to the rotary-wing UAVs, the greatest advantages are vertical flight capability, which facilitates takeoffs and landings, and their hovering capability, which allows them to fly in narrow space and around obstacles. But due to the short flight endurance and low speed, more flights may be necessary. There are also hybrid constructions, which can be fast and stable, have long flight endurance and allow vertical takeoffs and landings.

Flight Control System: The flight control system may be seen as the brain of the UAV, which is responsible for important tasks such as attitude control, flight path control, load equipment control, and fault detection. The flight of UAVs may be under remote control by an operator, or with various degrees of autonomy, such as autopilot assistance, up to fully autonomous aircraft that have no provision for human intervention. Advanced UAVs use GNSS, inertial navigation system (INS), geomagnetic sensors and barometers to allow for smarter navigation. With the signals received from orbiting satellites, UAVs can calculate their positions and navigate in real time. However, the satellite signals may be blocked by natural valley and man-made structures in the city, and a common method to overcome this challenge is the combination of GNSS and INS. These inertial measurement units (IMUs) provide linear acceleration across three axes, as well as pitch, roll, and yaw. Since the inertial sensors do gradually accumulate errors over time, the data fusion algorithm of GNSS and INS picks the most accurate estimate of position.

Ground Control System: The ground control system or ground control station (GCS) is a command center, which is responsible for UAV's flight control, airborne equipment control, flight status monitoring, and data link management. The GCS mainly consists of a computer, route planning software and flight monitoring software. The main function of the route planning software is to determine the flight path and the working position of the load sensor, in which image overlap, image resolution, sensor parameters, and the terrain should be considered. The flight monitoring software is used to provide flight parameters and navigation information in digital and graphic forms, and to assess the validity and quality of data collected by sensor systems.

Sensor System: As the applications of UAVs are expanding to various science and engineering disciplines, UAVs are often equipped with various sensors, including optical camera, thermal sensors, multispectral camera, LiDAR, SAR, and POS, increasing monitoring, feature tracking and object detection abilities.[③] Digital cameras are main sensors for image acquisition with centimeter-level resolution, which are suitable for large scale topographic mapping. When equipped with oblique cameras, building information can be obtained from multiple angles and used to create 3D real scene for smart city construction. LiDAR has the advantages of high ranging precision, large point density and strong penetration ability, but such sensors are at relatively high cost and heavy weight. LiDAR point cloud is usually used for DEM and DSM generation, ground surface classification, and 3D modeling. Based on D-GNSS and INS, POS is integrated with above-mentioned sensors to obtain the spatial position and attitude of these sensors.

In the future, the miniaturization, light weight, and deep integration of load sensors will become a major direction in UAV mapping.

Data Links: The data links are important part in the UAS responsible for data transmission between UAVs and the GCS. The transmitted data include remote command, flight parameters, images and point cloud acquired by airborne sensors, and so on. With the rapid development of 5G and satellite communication technology, it will play an important role in long-distance remote control and real-time transmission of large-bandwidth data for UAV mapping.

Characteristics of UAV Mapping

Compared to the traditional mapping methods, UAV mapping technology has the following advantages.

1) High flexibility. UAVs usually fly at low altitudes, so it is convenient to apply for airspace. Moreover, the lightweight UAVs can take off and land without professional airports, that is, they can easily take off on roads or hard grass in the mapping of difficult terrain.
2) High resolution. The spatial resolution of civil satellite images is 30cm and that of traditional aerial images is about 10cm, while the resolution of low-altitude UAV images can reach 2cm.
3) High quality. Since UAVs can fly low below clouds, images can be obtained without the effect of clouds, breaking the imaging limitations of weather conditions on high-altitude remote sensing.
4) High efficiency and low cost. Traditional mapping methods require detailed preparation, planning, and measurement, whereas UAVs can spend less time in automated data acquisition. Once the route is planned, data can be quickly collected by UAV sensors, at least covering dozens of square kilometers per day. Sensors on UAVs are generally light and small digital cameras, LiDAR, and IMUs with relatively low cost.
5) Safety. UAV mapping technology enables easy data collection without posing the risk of rough terrain, crumbling structures, high locations, and areas effected by natural disasters to surveyors.

Due to the limitations on the performance of UAVs in terms of payload, endurance, control, and wind resistance, there are certain limitations in UAV mapping. A UAS will not eliminate the need for surveyors on the ground or traditional mapping methods. For example, control points survey will determine the overall accuracy of the UAV mapping, and the horizontal and vertical errors of UAV mapping cannot be smaller than the site control points.

Applications

With numerous technical advantages, UAV mapping is widely used in large-scale mapping based on stereophotogrammetry, oblique photogrammetry and LiDAR. Among them, UAV aerial photogrammetry is mainly used in the production of DEMs and DSMs, DOMs, and DLGs at 1∶500-1∶2000 scale. With POS-assisted aerial triangulation technology, it can meet the mapping requirements of aerial survey at 1∶500 scale without control points. Compared to traditional vertical photogrammetry, UAV oblique photogrammetry can significantly improve the accuracy of elevation measurement, so it gradually becomes the main data source for 3D terrain reconstruction. With the rapid development of UAVs equipped with LiDAR, they are widely used in high-precision mapping, especially in mountainous areas covered by forests.

Multi-perspective UAV images with high resolution, realistic texture, and rich details can be quickly acquired, which have become an important data source for the national 3D real scene construction. Note that the 3D model based on UAV oblique photogrammetry is generally a continuous TIN mesh model, and the monomer model production is still challenging.④ Besides, the high-resolution images and high-density laser data can provide rich scene and target information, facilitating high-precision scene perception and target recognition.

UAV mapping is frequently used in disaster relief for its high flexibility and efficiency. The quick acquisition of images and 3D modeling for affected areas is beneficial to disaster assessment, search and rescue, and relief supplies. UAV remote sensing can also be used to monitor crop growth and strengthen field management, which is a typical application in precision agriculture to improve crop yield. UAV-based LiDAR can obtain better topographic and vegetation data with its ability of vegetation penetration and multiple echoes, to provide accurate information for forestry surveys. Other contribution of UAV mapping includes but are not limited to mine survey, power line inspection, marine monitoring, archaeological sites monitoring.

Words and Expressions

aerodynamic [ˌeərəʊdaɪˈnæmɪk] adj.空气动力学的
lift [lɪft] n.(空气的)升力;电梯;举;抬;提 v.举起;提高;解除;消散
turnaround [ˈtɜːnəraʊnd] n.周转;转变;转向;好转
industrialization [ɪndʌstrɪəlaɪˈzeɪʃn] n.产业化;工业化
stitch [stɪtʃ] v.缝;缝补;缝合 n.针脚;缝线;缝纫法
composite [ˈkɒmpəzɪt] adj.合成的;混合的;复合的 n.合成物;混合物;复合材料
aviation [ˌeɪviˈeɪʃn] n.航空;航空工业;飞行(术)

multirotor [ˌmʌltiˈroʊtər] *adj.*有多个转子的；有多旋翼的 *n.*多轴飞行器
hybrid [ˈhaɪbrɪd] *n.*混合物 *adj.*混合的；杂交成的
endurance [ɪnˈdʒʊərəns] *n.*忍耐力；耐久性
obstacle [ˈɒbstəkl] *n.*障碍；阻碍；障碍物
intervention [ˌɪntəˈvenʃn] *n.*干预；介入；调解
geomagnetic [ˌdʒiːəʊmæɡˈnetɪk] *adj.*地磁的；地磁学的
block [blɒk] *v.*阻塞；阻碍；堵塞 *n.*块；块体；区；障碍
pitch [pɪtʃ] *n.*[航]俯仰；沥青；球场；场地；颠簸；倾斜度
roll [rəʊl] *n.*[航]翻滚；横滚；卷；滚动 *v.*(使)翻滚；滚动
yaw [jɔː] *n.*(火箭、飞机、宇宙飞船等)偏航
miniaturization [ˌmɪnətʃərAɪˈzeɪʃn] *n.*微型化；小型化
crumbling [ˈkrʌmblɪŋ] *adj.*破碎的；摇摇欲坠的
texture [ˈtekstʃər] *n.*纹理；质地；口感
monomer [ˈmɒnəʊmər] *n.*单体；单元结构
perception [pəˈsepʃn] *n.*感觉；感知；看法；洞察力

☞ **Notes**：

①无人机测绘技术指的是通过搭载的各种传感器(如高分辨率数码相机、激光扫描仪)进行数据获取，用相应软件进行数据处理，并且按一定精度要求进行成果输出的技术手段。
②无人机可以是固定翼的、旋翼或多旋翼的，以及混合动力的。与旋翼无人机相比，固定翼无人机的机翼扁平、耗能更少，从而更适合长距离飞行。
③随着无人机的应用扩展到各个科学和工程学科领域，无人机通常会装备各种传感器，包括光学相机、热传感器、多光谱相机、激光雷达、合成孔径雷达以及定位定向系统(POS)，从而增加了监测、特征跟踪和目标检测的能力。
④无人机能够快速获取高分辨率、纹理真实、细节丰富的多视影像，已经成为实景三维国家建设的重要数据源。需要注意的是，基于无人机倾斜摄影测量建立的三维模型一般是连续的 TIN 网格模型，生成单体模型仍然具有挑战性。

Terms Highlights

flying platform 飞行平台
flight control system 飞行控制系统
fault detection 故障检测
inertial navigation system (INS) 惯性导航系统
pitch, roll, and yaw/heading 俯仰；横滚；航偏
POS-assisted aerial triangulation POS 辅助空中三角测量
ground control system, ground control station (GCS) 地面控制系统；地面控制站

sensor system 传感器系统
image overlap 影像重叠
sensor parameters 传感器参数

Part III

Basic International Academic Communication

Unit 27　International Geoscience Organizations

FIG — International Federation of Surveyors

FIG was founded in 1878 in Paris. It is a federation of national associations and is the only international body that represents all surveying disciplines. It is an UN-recognized non-government organization (NGO) and its aim is to ensure that the disciplines of surveying and all who practice them meet the needs of the markets and communities that they serve. It realizes its aim by promoting the practice of the profession and encouraging the development of professional standards. FIG's technical work is led by ten commissions. The commissions' activity is to prepare and conduct the program for FIG's international congresses, held every four years, and annual working weeks, held in the intervening years. 10 commissions are:

1 — Professional Practice
2 — Professional Education
3 — Spatial Information Management
4 — Hydrography
5 — Positioning and Measurement
6 — Engineering Surveys
7 — Cadastre and Land Management
8 — Spatial Planning and Development
9 — Valuation and the Management of Real Estate
10 — Construction Economics and Management

IUGG — International Union of Geodesy and Geophysics

The International Union of Geodesy and Geophysics (IUGG) is a scientific organization dedicated to promoting and coordinating studies of the Earth and its environment in space. It is a non-governmental, scientific organization, established in 1919. IUGG is one of the 30 scientific Unions presently grouped within the International Council for Science.

IUGG is dedicated to the international promotion and coordination of scientific studies of Earth (physical, chemical, and mathematical) and its environment in space. These studies include the shape of the Earth, its gravitational and magnetic fields, the dynamics of the Earth as a whole and of its component parts, the Earth's internal structure, composition and tectonics, the generation of magmas, volcanism and rock formation, the hydrological cycle including snow and ice, all aspects of the oceans, the atmosphere, ionosphere, magnetosphere and solar-terrestrial relations, and analogous problems associated with the Moon and other planets. IUGG encourages the application of this knowledge to societal needs, such as mineral resources, mitigation of natural hazards and environmental preservation.

IUGG is comprised of eight semi-autonomous associations, each responsible for a specific range of topics or themes within the overall scope of Union activities. In addition, IUGG establishes Inter-Association Commissions, and relationships with several other scientific bodies with similar interests. The International Association of Geodesy (IAG) is one of seven associations within the International Union of Geodesy and Geophysics.

IUGG holds General Assemblies at four-year intervals, and each of its associations organize scientific assemblies as well as topical symposia in the intervening period between General Assemblies. IUGG is committed to the principle of free exchange of data and knowledge among nations, and encourages unreserved scientific participation by all peoples.

IAG-International Association of Geodesy

The International Association of Geodesy, hereafter called the Association or the IAG, is a constituent Association of the International Union of Geodesy and Geophysics, hereafter called the Union or IUGG, and is subject to the Statutes and Bylaws of the Union as well as to these statutes.

The Mission of IAG is the advancement of geodesy. It is implemented by furthering geodetic theory through research and teaching, by collecting, analyzing, modeling and interpreting observational data, by stimulating technological development and by providing a consistent representation of the figure, rotation, and gravity field of the Earth and planets, and their temporal variations. The objectives cover the study of all geodetic problems related to Earth observation and global change. This comprises the establishment of reference systems, monitoring the gravity field and rotation of the Earth, the deformation of the Earth's surface including ocean and ice, and positioning for interdisciplinary use. IAG shall initiate, coordinate, and promote international cooperation and knowledge exchange through symposia, workshops, summer schools, publications, and other means of communication. The goal is to foster the development of geodetic activities and infrastructure in all regions of the world, taking into consideration the specific situation of developing countries. The structure includes the

following components:
- Commission 1: Reference Frames
- Commission 2: Gravity Field
- Commission 3: Geodynamics and Earth Rotation
- Commission 4: Positioning and Applications
- Inter-commission Committee on Theory
- Global Geodetic Observing System (GGOS)
- 15 International Scientific Services

ISPRS—International Society for Photogrammetry and Remote Sensing

The International Society for Photogrammetry and Remote Sensing (ISPRS) is a non-governmental organization devoted to the development of international cooperation for the advancement of photogrammetry and remote sensing and their applications. The Society operates without any discrimination on grounds of race, religion, nationality, or political philosophy. The official languages are English, French and German.

The Society's scientific interests include photogrammetry, remote sensing, spatial information systems and related disciplines, as well as applications in cartography, geodesy, surveying, natural, Earth and engineering sciences, and environmental monitoring and protection. Further applications include industrial design and manufacturing, architecture and monument preservation, medicine and others. Seven commissions are:

1—Image Data Acquisition-Sensors and Platforms
2—Theory and Concepts of Spatial Information Science
3—Photogrammetric Computer Vision and Image Analysis
4—Geodatabases and Digital Mapping
5—Close-Range Sensing: Analysis and Applications
6—Education and Outreach
7—Thematic Processing, Modeling and Analysis of Remotely Sensed Data
8—Remote Sensing Applications and Policies

International Cartographic Association

ICA is the world authoritative body for cartography, the discipline dealing with the conception, production, dissemination and study of maps. The mission of the International Cartographic Association is to promote the discipline and profession of cartography in an international context.

International Hydrographic Organization

The International Hydrographic Organization is an intergovernmental consultative and technical organization that was established in 1921 to support the safety in navigation and the protection of the marine environment.

Words and Expressions

federation [ˌfedəˈreɪʃn] n.联合；同盟；联邦；联盟
community [kəˈmjuːnəti] n.团体；社会；(政治)共同体
commission [kəˈmɪʃn] n.委员会
congress [ˈkɒŋɡres] n.(代表)大会；[C-](美国及其他国家的)国会；议会
intervene [ˌɪntəˈviːn] v.干涉；干预；插入；介入；(指时间)介于其间
Munich [ˈmjuːnɪk] n.慕尼黑(德国城市；巴伐利亚州首府)
valuation [ˌvæljuˈeɪʃn] n.估价；评价；计算
real estate [ˈriːəl ɪsˈteɪt] n.房地产；房地产所有权
dedicate [ˈdedɪkeɪt] vt.献(身)；致力；题献
council [ˈkaʊnsl] n.理事会；委员会；参议会；讨论会议；顾问班子；立法班子
dynamics [daɪˈnæmɪks] n.动力学
internal [ɪnˈtɜːnl] adj.内在的；国内的
tectonics [tɛkˈtɒnɪks] n.构造学；筑造学
magma [ˈmæɡmə] n.岩浆；[药]乳浆剂
volcanism [ˈvɒlkənɪzəm] n.火山作用
magnetosphere [ˈmæɡnɪtəʊsfɪər] n.磁气圈
solar-terrestrial relation 日地关系
analogous [əˈnæləɡəs] adj.类似的；相似的；可比拟的
mitigation [ˌmɪtɪˈɡeɪʃn] n.缓解；减轻；平静
hazard [ˈhæzəd] n.冒险；危险；冒险的事
preservation [ˌprezərˈveɪʃn] n.保存
autonomous [ɔːˈtɒnəməs] adj.自治的
association [əˌsəʊsɪˈeɪʃn] n.协会；联合；结交；联想
assembly [əˈsembli] n.集合；装配；集会；集结；汇编
symposia [sɪmˈpəʊziə] n.座谈会；评论集
unreserved [ˌʌnrɪˈzɜːvd] adj.坦白的；不隐瞒的；无限制的
constituent [kənˈstɪtjuənt] adj.有选举权的；有宪法制定权的；组成的
statute [ˈstætʃuːt] n.法令；条例
bylaw [ˈbaɪlɔː] n.次要法规；(社、团制定的)规章制度
outreach [ˈaʊtriːtʃ] v.到达顶端；超越

branch [brɑːntʃ] n.分部；分店；分科；部门；支流
discrimination [dɪˌskrɪmɪˈneɪʃn] n.辨别；区别；识别力；辨别力；歧视
philosophy [fəˈlɒsəfi] n.哲学；哲学体系；达观；冷静

Terms Highlights

UN(United Nations) 联合国

FIG(Federation of International Surveyors) 国际测量师联合会

IUGG(International Union of Geodesy and Geophysics) 国际大地测量与地球物理联合会

IAG (International Association of Geodesy) 国际大地测量协会

ISPRS(International Society for Photogrammetry and Remote Sensing) 国际摄影测量与遥感学会

ICA(International Cartographic Association) 国际制图协会

IHO(International Hydrography Organization) 国际海道测量组织

IGU (International Geographical Union) 国际地理联合会

IAM(International Society of Mine Surveying) 国际矿山测量协会

ISDE (International Society for Digital Earth) 国际数字地球协会

CSGPC(Chinese Society of Geodesy, Photogrammetry and Cartography) 中国测绘学会

Unit 28　Prestigious Journals in Geomatics

Journal of Geodesy

The Journal of Geodesy is an international journal concerned with the study of scientific problems of geodesy and related interdisciplinary sciences. Peer-reviewed papers are published on theoretical or modeling studies, and on results of experiments and interpretations. Besides original research papers, the journal includes commissioned review papers on topical subjects and special issues arising from chosen scientific symposia or workshops. The journal covers the whole range of geodetic science and reports on theoretical and applied studies in research areas such as: Positioning; Reference frame; Geodetic networks; Modeling and quality control; Space geodesy; Remote sensing; Gravity fields; Geodynamics.

　　ISSN: 0949-7714
　　Frequency: monthly
　　Website: http://link.springer.de/link/service/journals/00190/index.htm

Journal of Surveying Engineering

The Journal of Surveying Engineering covers the broad spectrum of surveying and mapping activities encountered in modern practice. It includes traditional areas such as construction surveys, control surveys, photogrammetric mapping, engineering layout, deformation measurements, precise alignment, and boundary surveying. It also includes newer development such as satellite positioning; spatial database design, quality assurance, and information management of geographic information systems; computer applications involving modeling, data structures, algorithms, and information processing; digital mapping, coordinate systems, cartographic representations, and the role of surveying engineering professionals in an information society.

　　ISSN: 0733-9453
　　Frequency: quarterly
　　Website: http://scitation.aip.org/suo/

Survey Review

Survey Review, was first published in 1931 as a result of a resolution of the First Commonwealth Survey Officers Conference in 1928. It has been published continuously since then as a quarterly journal, bringing together a wide range of papers on research, theory, practice and management in land and engineering surveying. All papers are independently assessed by two referees and come from government, private industry and academic organizations worldwide.

Survey Review is included in the Institute of Scientific Research (ISR) index of the most important and influential research conducted throughout the world. It is published quarterly, in January, April, July and October, by the Commonwealth Association for Surveying and Land Economy (CASLE).

ISSN: 0039-6265
Frequency: quarterly
Website: http://www.surveyreview.org/index.html

Journal of Geodynamics

The Journal of Geodynamics is an international and interdisciplinary forum for the publication of results and discussions of solid earth research in geodetic, geophysical, geological and geochemical geodynamics, with special emphasis on the large scale processes involved. Papers addressing interdisciplinary aspects, analyses, results and interpretation will receive special attention. Original research papers, including "letters", as well as topical reviews are invited.

The emphasis lies on endogenic and also interacting exogenic processes and their geological effects in the widest sense, as well as on results obtained from geophysical, geodetic and geological measurements and analytical techniques applied to deduce them.

ISSN: 0264-3707
Frequency: 10 issues per year
Website: http://www.elsevier.com/wps/find/journaldescription.cws_home/874/description

Journal of Geophysical Research

American Geophysical Union (AGU) is a worldwide scientific community that advances, through unselfish cooperation in research, the understanding of Earth and space for the benefit of humanity. AGU currently publishes a weekly newspaper, twelve scientific journals, and nine book series. Making the results of geophysical research available to the individuals who need them to advance the understanding of Earth and space is one of the primary purposes of the organization.

The Journal of Geophysical Research is one of twelve scientific journals published by (AGU) which divide into six aspects: JGR-Atmospheres; JGR-Earth Surface, JGR-Oceans; JGR-Planets; JGR-Solid Earth; JGR-Space Physics.

JGR-Solid Earth focuses on the physics and chemistry of the solid Earth and the liquid core of the Earth, geomagnetism, paleomagnetism, marine geology/geophysics, chemistry and physics of minerals, rocks, volcanology, seismology, geodesy, gravity, and tectonophysics.

ISSN: 0148-0227

Frequency: monthly

Website: http://www.agu.org/pubs/pubs.html#journals

Impact Factor 2.245. Ranked #12 of 122 journals in the Geosciences, Interdisciplinary Category of the 2002 Journal Citation Reports.

Geophysical Research Letters

Geophysical Research Letters publishes short, concise research letters that present scientific advances that are likely to have immediate influence on the research of other investigators. *GRL* can focus on a specific discipline or apply broadly to the geophysical science community.

GRL is a letters journal that is also published by AGU; limiting manuscript size expedites the review and publication process. A magazine-sized *GRL* can reach and be of interest to the largest AGU audience. With this goal, the Editorial Board evaluates manuscripts submitted to *GRL* according to the following criteria: (1) Is it a short, concise research letter? (2) Does it contain important scientific advances? (3) Would it have immediate impact on the research of others?

GRL has been in publication since 1974. *The Editors* are adapting *GRL* to the evolving needs of the Earth science community. *GRL Editors* are now topical: the subject of the paper determines which *Editor* handles the review.

ISSN: 0094-8276

Frequency: semimonthly

Website: http://www.agu.org/journals/gl/.

Impact Factor 2.150. Ranked #13 of 122 titles in Geosciences in the 2002 Journal Citation Reports.

GPS Solutions

GPS Solutions is a scientific journal. It is published quarterly and features system design issues and a full range of current and emerging applications of global navigation satellite systems (GNSS) such as GPS, GLONASS, Galileo and various augmentations.

Novel, innovative, or highly demanding uses are of prime interest. Areas of application include: aviation, surveying and mapping, forestry and agriculture, maritime and waterway navigation, public transportation, time and frequency comparisons and dissemination, space and satellite operations, law enforcement and public safety, communications, meteorology and atmospheric science, geosciences, monitoring global change, technology and engineering, GIS, geodesy, and others.

ISSN: 1080-5370

Frequency: quarterly

Website: link.springer.de/link/service/journals/10291/

GPS World

Through its tutorial columns, case studies, product notes, and industry news, *GPS World* provides practical, cutting-edge information for engineers, systems integrators, end users, and consultants. *GPS World* enables its readers to more effectively meet the development challenges of location-sensitive products and applications that incorporate Global Positioning Systems (GPS), sensors, wireless communications, mobile computing, the Internet, and other technologies. When appropriate, *GPS World's* editorial content is presented within the context of operational requirements, policy issues, systems modernization efforts, and relevant science.

Founded in 1989, *GPS World* is the first and only international monthly journal that features news and applications of the Global Positioning System, the developing Galileo system, Glonass, and related technologies.

ISSN: 1048-5104

Frequency: monthly

Website: http://www.gpsworld.com/gpsworld/

International Journal of Geographical Information Science

The aim of *This Journal* is to provide a forum for the exchange of original ideas, techniques, approaches and experiences in the rapidly growing field of geographical information systems (GIS). It is intended to interest those who design, implement and use systems for monitoring, planning and policy-making. Published research covers applications of GIS in natural resources, social, systems, and the built environment, and developments in computer science, cartography, surveying, geography and engineering in both developed and developing countries.

ISSN: 1365-8816

Frequency: 8 issues per year

Website: http://www.tandf.co.uk/journals/titles/13658816.asp

ISPRS Journal of Photogrammetry and Remote Sensing (P&RS)

ISPRS Journal of Photogrammetry and Remote Sensing (P&RS) is the official journal of the International Society for Photogrammetry and Remote Sensing (ISPRS). *The Journal* is to provide a channel of communication for scientists and professionals in all countries working in the many disciplines employing photogrammetry, remote sensing, spatial information systems, computer vision, and other related fields. *The Journal* is designed to serve as a source reference and archive of advancements in these disciplines.

The P & RS objective is to publish high quality, peer-reviewed, preferably previously unpublished papers of a scientific/research, technological development or application/practical nature. *The P & RS* will publish papers, including those based on ISPRS meeting presentations, which are regarded as significant contributions in the above-mentioned fields.

ISSN: 0924-2716

Frequency: bimonthly

Website: http://www.sciencedirect.com/science/journal/09242716

Geomatica

Geomatica (formerly *CISM Journal ACSGC*) is the official quarterly publication of the Canadian Institute of Geomatics. It is the oldest surveying and mapping publication in Canada and was first published in 1922 as the Journal of the Dominion Land Surveyors' Association.

Geomatica is dedicated to the dissemination of information on technical advances in the geomatics sciences. The internationally respected publication contains special features, notices of conferences, calendar of event, articles on personalities, review of current books, industry news and new products, all of which keep the publication lively and informative. It provides a useful mix of practical, professional, and academic articles with opinions and information from Canada and abroad.

ISSN: 1195-1036

Frequency: quarterly

Website: http://www.cig-acsg.ca

Other Related Magazines:

GeoInformatica: An International Journal on Advances of Computer Science for Geographic Information Systems (quarterly) http://www.kluweronline.com/issn/1384-6175/contents

Marine Geodesy: An International Journal of Ocean Surveys, Mapping, and Sensing (4 issues per year) http://www.tandf.co.uk/journals/titles/01490419.asp

Geospatial Solutions: http://www.geospatial-online.com/geospatialsolutions/
   Photogrammetric Engineering and Remote Sensing http://www.geo-online.org/index.shtml

Words and Expressions

prestigious [preˈstɪdʒəs] *adj.* 享有声望的；声望很高的
interdisciplinary [ˌɪntəˈdɪsəplɪnəri] *adj.* 各学科间的
workshop [ˈwɜːkʃɒp] *n.* 研讨会；车间
ISSN *abbr.* 国际标准期刊编号
commonwealth [ˈkɒmənwelθ] *n.* 国民整体；共和国；联邦
referee [ˌrefəˈriː] *n.* 仲裁人；调解人；[体]裁判员
forum [ˈfɔːrəm] *n.* 论坛；法庭；讨论会
endogenic [ˌendəʊˈdʒenɪk] *adj.* [地]内成的；内生的；内营力的
exogenic [ˌeksəʊˈdʒenɪk] *adj.* [生]外生的；[地]外成的；[医]外因的
geomagnetism [ˌdʒiːəʊˈmæɡnətɪzəm] *n.* 地磁；地磁学
paleomagnetism [ˌpæliəʊˈmæɡnətɪzəm] *n.* [物]古地磁学
volcanology [ˌvɒlkəˈnɒlədʒi] *n.* 火山学
seismology [saɪzˈmɒlədʒi] *n.* 地震学
tectonophysics [ˌtektəʊnəʊˈfɪzɪks] *n.* [物]地壳构造物理学
semimonthly [ˌsemɪˈmʌnθli] *adj.* 每半个月一次的 *n.* 半月刊
tutorial [ˈtjuːtərɪəl] *n.* 指南 *adj.* 导师的
consultant [kənˈsʌltənt] *n.* 顾问；商议者；咨询者
dissemination [dɪˌsemɪˈneɪʃn] *n.* 分发

Terms Highlights

Journal of Geodesy 大地测量杂志
Journal of Surveying Engineering 测量工程杂志
Survey Review 测量评论
Journal of Geodynamics 地球动力学杂志
Journal of Geophysical Research 地球物理学研究杂志
Geophysical Research Letters 地球物理研究快报
GPS Solution 全球定位系统
GPS World 全球定位系统世界
International Journal of Geographical Information Science 国际地理信息系统杂志
ISPRS Journal of Photogrammetry and Remote Sensing (*P & RS*) 国际摄影测量与遥感

学会杂志

Geomatica 地球空间信息学
GeoInformatica 地学信息
Marine Geodesy 海洋大地测量学
Geospatial Solutions 地球空间情报
Photogrammetric Engineering and Remote Sensing 摄影测量工程与遥感
Cartography and Geographic Information System 制图学和地理信息系统
The Cartographic Journal 制图学杂志
Computers & Geosciences 计算机与地学
IEEE Transactions on Geoscience & Remote Sensing IEEE 地球科学与遥感汇刊
Remote Sensing of Environment 环境遥感
International Journal of Remote Sensing 国际遥感杂志

Unit 29 Professional English Paper Writing

Types of Professional Papers

A professional paper is a formal printed document in which professionals present their views and research findings on any deliberately chosen topic. Professional papers assigned in universities and colleges or any other research institutions are generally of the following four types: (1) report paper; (2) research paper; (3) term paper, and (4) thesis paper (dissertation).

Report Paper: A report paper refers to an account of some particular subject or issue which is published in a newspaper or magazine or broadcast over the radio or on the television. It summarizes and reports the findings of the author(s) on a particular subject. The author(s) may not give his/their own opinion on the issue, nor evaluates the findings, but merely catalogs them in a sensible sequence.

Research Paper: A research paper is also known as a library paper, or a thesis, which refers to the formal, fairly long and well-documented composition which aims to explore, discuss, analyze and finally to find new opinions and results in a certain factual or theoretical issue. It usually deals with a relatively specific topic, and a comparatively narrow subject.

Term Paper: A term paper refers to a major assignment in a school or college course representative of a student's achievement or achievements during a particular term, also known as a course paper.

Thesis (dissertation) paper: Thesis as a term may have different meanings. A thesis refers to an idea or theory that is expressed as a statement to show its author's standpoint or belief on some particular issue. However, the term thesis can also be used to suggest a research paper with equal value or formality to a degree paper, though the form may usually be briefer or less exhaustive than a treatise or dissertation. When thesis is used to replace dissertation, it gives the listener or reader a sense of informality.

Structural Features of Research Paper

A complete research paper is usually composed of the following elements: title, author, affiliation, abstract, keywords, introduction, theoretical analysis and/or experimental description, results and discussion or conclusion, acknowledgments, references, etc.

Title: A title is the generalization of a text. It should summarize the central idea of the paper concisely and correctly, attract the reader and facilitate the retrieval. The words or phrases used in a title are very often nouns, noun phrases or gerunds which usually are keywords for the paper.

Author and affiliation: The author is the person who performs the entire work of the paper writing and is responsible for the content of the paper. When there is more than one author, the first person is called principal author. The names of author(s) and affiliations are usually used as author indexing, which can be easily corresponded with the author, the author's institution and his country.

Abstract and keywords: An abstract is a condensed statement of the contents of a paper. As a short, concise and highly generalized text, an abstract is viewed as a mini-version or a miniature of the document, summarizing the content of the main body with a strictly limited number of words. Normally, 200 words should be a sensible maximum for a relatively long paper but never more than 500 words. As a general rule, an abstract will approximately 3% ~5% of the length of the paper, but is seldom more than 2/3rds of a page. Keywords are the most important words and phases to represent the theme or subject of the paper. In general, the number of keywords should be at least 2 and at most 8.

Introduction: Every professional paper should have at least one or two introductory paragraphs with or without a particular subtitle. The length or the degree of formality of a paper may decide whether the introduction should be a separate-labeled section. Generally speaking, the successful introduction of a paper should have the following four functions to facilitate the communication process: (1) introducing the subject to supply sufficient background information to the readers; (2) limiting the research scope and narrowing down the scope of the work; (3) stating the general purpose of the paper and illustrate the primary objectives of his research; and (4) showing the writing arrangement which can usually be found at the end of the introduction.

Theoretical analysis and/or experimental description: No professional paper is purely theoretical or experimental. General speaking, a theoretic analysis which consists of principle used, mathematical model and equations should be proved by experimental description or practical applications. Likewise, the experimental description is usually based on some mathematical model or theoretical analysis. Sometimes the experimental description can also lead to a new explanation, discovery or development of the theoretical model.

Results and discussion or conclusion: The value of a research lies in the value of its final results and the author's interpretation of the results. They usually presented together with the corresponding analysis concerned. So in this section the author should present the essential results and data and then generalize them to a theoretical height.

Acknowledgments: Acknowledgment(s) is mainly used to extend the author's indebtedness to the helpful support or concern from his colleagues or institution in offering any useful support etc. to the author.

References: Reference or bibliography of a paper may is a list of books and papers cited in a research paper. The Harvard system is generally used which is alphabetically arranged and headed by the author's name, the year of publication, the title of the publication, the publisher, the page number and so on.

An Example of a Research Paper
(Selected from GEOSPATIAL INFORMATION SCIENCE Vol.7 2004(4), pp279~283)

Least Squares Spectral Analysis and Its Application to Superconducting Gravimeter Data Analysis

Yin Hui[1] and Spiros D. Pagiatakis[2]

1 School of Geodesy and Geomatics, Wuhan University, 129 Luoyu Road, Wuhan, 430079, China
2 Department of Earth and Space Science and Engineering, York University, Toronto, Canada.

Abstract

Detection of a periodic signal hidden in noise is the goal of Superconducting Gravimeter (SG) data analysis. Due to spikes, gaps, datum shrifts (offsets) and other disturbances, the traditional FFT method shows inherent limitations. Instead, the least squares spectral analysis (LSSA) has been shown to be more suitable than Fourier analysis of gappy, unequally spaced and unequally weighted data series in a variety of applications in geodesy and geophysics. This paper reviews the principle of LSSA and gives a possible strategy for the analysis of time series obtained from the Canadian Superconducting Gravimeter Installation (CGSI), with gaps, offsets, unequal sampling decimation of the data and unequally weighted data points.

Key words: Least Squares Spectrum, Superconducting Gravimeter, Data Analysis

1. Introduction

The fast Fourier transform (FFT) algorithms are popular spectral estimation techniques for the determination of the power spectrum and show computational efficiency in the analysis of signal process. However, there are certain inherent limitations in the FFT techniques. The most prominent limitations arise from the requirement that the data be equally spaced and equally weighted with no gaps and datum shifts. Pre-processing of the data is required if there are gaps, spikes, datum shifts and trends in the original data series.

In order to avoid unnecessary data pre-processing that may corrupt or obliterate the useful information hidden in the series (signal) (Pagiatakis, 2000), the Least Squares Spectral Analysis (LSSA) is used as an alternative to the classical Fourier method. LSSA was first developed by Vanicek (1969, 1971). It uses the least squares approximation (LSA) technique [Vanicek & Wells, 1972], which is closely related to the least squares parametric adjustment (LLSPA) [Wells & Krakiwsky, 1971; Vanicek & Krakiwsky, 1982]. The advantages of this technique have already been shown by Pagiatakis, (1999).

The principle of LSSA is introduced in this paper based on the least squares parametric adjustment in vector space. The experimental studies of this method are conducted using the observation series from the Canadian Superconducting Gravimeter Installation (CSGI), which exhibit minor gaps (unequally spaced), offsets, spikes (seismic events) and variable weights due to different noise levels. This research work was carried out by the principal author as a visiting scholar at York University, Toronto, Canada, in 2003. It is also part of the research carried out for the Global Geodynamics Project (GGP). The general procedures of data analysis by LSSA are also given in this paper. Finally, several useful conclusions are drawn.

2. Principle of LSSA

2.1 Least squares parametric adjustment in vector space

Before describing LSSA, it is expedient to refer to the least squares parametric adjustment in a vector space (Wells et al., 1985; Chen, 1988). As we know, the shortest distance between a point and a plane is the perpendicular from the point to the plane (projection theorem). We can also extend this notion to n-dimensional space, in which the shortest distance between a point and any sub-space is the perpendicular from the point to the sub-space. Below is the mathematical description of the projection theorem.

2.1.1 Projection theorem

Let H be a vector space with inner product, called Hilbert space. Given $f \in H$ (point) and $M \in H$ (sub-space), then of all the elements $m \in M$, there is one element $m_0 \in M$ such that $\|f-m_0\| < \|f-m\|$. This element m_0 is given by the orthogonal projection of f onto M, that is $f-m_0 \perp M$ (see Figure 1).

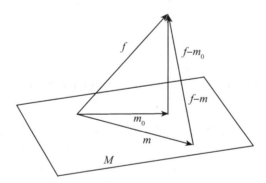

Figure 1   Projection

If M is generated by a set of basis vectors $\{\varphi_i\}$, then m_0 can be expressed as a linear combination of these basis vectors

$$m_0 = \sum_i x_i \varphi_i \tag{1}$$

Since $(f-m_0) \perp M$, we have $(f-m_0) \perp \varphi_j$, $\forall j$, therefore

$$\langle f - \sum_i x_i \varphi_i, \varphi_j \rangle = 0, \forall j. \tag{2}$$

By using the inner product, we get

$$\sum_i \langle \varphi_i, \varphi_j \rangle x_i = \langle f, \varphi_j \rangle, \quad \forall j. \tag{3}$$

In fact, this is the normal equation of parametric adjustment in which x_i is the unknown vector.

2.1.2 Least squares parametric adjustment in vector space

We can now use the concept of Hilbert space to the least squares parametric adjustment. Based upon the Gauss-Markov model $(l, AX, \sigma_0^2 I)$, l is a vector of observations, $(l \in H)$. The least squares principle aims to find a vector $\hat{l} = A\hat{X}$ to make the distance between l and $\hat{l}$ shortest or the norm of the residual vector v minimum. Therefore, from the projection theorem, $\hat{l}$ is the orthogonal projection of l onto a sub-space ($M \in H$). So from the above equation, we get

$$\hat{X} = (A^T A)^{-1} A^T l \tag{4}$$

$$v = l - \hat{l} = l - A(A^T A)^{-1} A^T l \tag{5}$$

From the projection theorem, we know that $v \perp \hat{l}$. This means that the projection theorem allows us to decompose l into two orthogonal components, $\hat{l}$ (the orthogonal projection of l onto M) and v (the perpendicular from l to M).

2.2 Least squares spectrum

Given a vector of observations $f = f(t) = \{f_i\}$, $i = 1, 2, \cdots, n$, we can set up a model p that can be expressed as follows.

$$p = \sum_i \varphi_i \hat{x} = \Phi \hat{X} \tag{6}$$

Where Φ is a matrix of known base functions and $\hat{X}$ is the vector of unknown parameters. Here we do not assume t_i to be equally spaced. But we assume that the observations f_i, $i = 1, 2, \cdots, n$ possess a fully populated covariance matrix C_f. For simplicity here we assume $C_f = I$.

To estimate the model parameters $\hat{X}$, the projection theorem is used and the difference between p and f becomes minimum in the least squares sense. The estimation of the model parameters can be obtained as follows:

$$\hat{X} = (\Phi^T \Phi)^{-1} \Phi^T f, \tag{7}$$

$$p = \Phi \hat{X} = \Phi (\Phi^T \Phi)^{-1} \Phi^T f, \tag{8}$$

The residual vector can be written as

$$v = f - p = f - \Phi(\Phi^T \Phi)^{-1} \Phi^T f, \tag{9}$$

From the projection theorem, we know that $v \perp p$. This means that f has been decomposed into a signal p and noise v (residual series). If we project p back onto f, we can get the length of this orthogonal projection as follows (see Figure 2).

$$\frac{\langle f, p \rangle}{\|f\|} \tag{10}$$

To describe how closely p approaches f, we use a fractional measure s that is the ratio of the length of this orthogonal projection to the length of f

$$s = \frac{\langle f, p \rangle}{\|f\| \, \|f\|} = \frac{f^T P}{f^T f}. \tag{11}$$

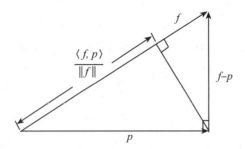

Figure 2 Second Projection

In spectral analysis we usually try to search for the hidden periodic signals that are expressed in terms of sine and cosine base functions. So, if we specify the form of the base functions to be trigonometric based on a set of spectral frequencies $\omega_i, i = 1, 2, \cdots, m$, we have

$$p(\omega_i) = \hat{X}_{1i}\cos\omega_i t + \hat{X}_{2i}\sin\omega_i t. \tag{12}$$

Let $\hat{X} = [\hat{X}_{1i}, \hat{X}_{2i}]^T$ and $\boldsymbol{\Phi} = [\cos\omega_i t, \sin\omega_i t]$, then $\hat{X}$ can be determined from (7). So for different frequencies $\omega_i, i = 1, 2, \cdots, m$, we can get different spectral values.

$$s(\omega_i) = \frac{f^T p(\omega_i)}{f^T f}, \quad i = 1, 2, \cdots, m. \tag{13}$$

Equation (13) describes the least squares spectrum. Obviously the least squares spectrum of f is the collection of the spectral values for all desired frequencies $\omega_i, i = 1, 2, \cdots, m$. The bigger the spectral value at a frequency ω_i, the more powerful f is at this frequency.

A least squares spectrum with a fully populated covariance matrix C_f can be expressed directly (Pagiatakis, 1999) as follows:

$$s = \frac{f^T C_f^{-1} \boldsymbol{\Phi} (\boldsymbol{\Phi}^T C_f^{-1} \boldsymbol{\Phi})^{-1} \boldsymbol{\Phi}^T C_f^{-1} f}{f^T C_f^{-1} f} \tag{14}$$

3. Superconducting Gravimeter Data Analysis

The Canadian Superconducting Gravimeter Installation (CSGI) is one of 20 superconducting gravimeters (SG) currently operating around the globe, under the auspices of an international program called the Global Geodynamics Project (GGP) (Crossley, 1994). The purpose of this program is to monitor changes in the Earth's gravity field at periods of seconds and longer. The observation period commenced in July 1997 and the first 6-year period was completed in June 2003. The measurements have been taken with the same formats and same intervals from all 20 SG installations. Below we give a possible strategy for the analysis of the CSGI data using LSSA.

3.1 Data acquisition

A six-year-long data series from the Canadian Superconducting Gravimeter Installation (CSGI) have been assembled for analysis. All raw data were digitally recorded at the sampling rate of 1-sec in units of Volt. Volts can be changed into units of gravity (μGal) by using a calibration factor usually obtained from parallel absolute gravity measurements. This calibration factor for CSGI is -78.48 μGal/V. Due to environmental disturbances and other processes (earthquakes and hydro-geological process) and instrumental factors (helium refills, failures, maintenance, calibration, system upgrades etc.), the SG data are not always continuous and quiet. Gaps, spikes and datum shifts as well as occasional high-level noise are present in the raw data.

3.2 Earth tide removal

A synthetic tide by the gravity tide prediction program G-WAVE was calculated at 1s intervals and removed from the superconducting gravimeter record. Program G-WAVE was written by J. B. Merriam (Merriam, 1992) at the University of Saskatchewan, Canada with predication accuracy of about 40 nGal, when over 3,000 tidal waves are used. The colatitude and longitude of the station are required to compute the gravity tide at different levels of accuracy depending on the number of waves used.

3.3 Data filtering

The 1s residual data series produce a very large data set and perhaps with a sampling interval that is unnecessarily short for our scope of analysis. Also, due to the limitation of the computer memory (desktop), the data series must be filtered in the time domain. A Parzen window is used as a weighting function throughout the whole series, purposely centered at random times with no overlap between successive windows to avoid correlation. The series produced by this approach is unequally spaced with variances derived from rigorous error propagation of the noise levels within the window to the filtered data value. To avoid any possibility of aliasing, approximately 24 points per hour with a window lag of 50s and randomly sampled with maximum step 100s were obtained. Figure 3 shows the filtered data series and some of the data sets. As an example, Table 1 (omit) shows a short segment of the series with unequally spaced and weighted values.

3.4 Least squares spectral analysis

After the initial processing of the data (unequally spaced decimation), we obtained filtered data sets with unequally spaced and unequally weighted values. From Figure 3 (omit), we can see clearly that there exists a sizeable noise in April 2001. The reason for this is a serious disturbance to the instrument. This high level noise will certainly distort the spectrum of the data series. Therefore, we eliminated these data segments leaving a gap. The left panel in Figure 4 (omit) shows the segment in April 2001 with the said gap and the right panel shows a data segment in June 2001 with datum shrifts and spikes. Based on the technique of LSSA, we did not need to take any corrective action (e.g., interpolation between gaps, smooth of spikes and datum shifts). The spectra were obtained by the weighted least-squares spectral analysis method. Each spectrum corresponds to a time series of one month. For example, the least squares spectra of April 2001 and June 2001 are shown in Figure 5(omit).

4. Discussions and Conclusion

The superconducting gravimeter time series contain a wealth of information about the internal process of the Earth that may be concealed by high frequency noise or may be destroyed by careless and unnecessary preprocessing of the original data. The experimental studies with LSSA not only do they show a great advantage when analyzing gappy, unequally spaced and unequally weighted data, but also allow us to draw a few useful conclusions as follows:

1) Data filtering can be done at unequal space intervals prior to their analysis by the LSSA. The so produced unequally spaced series values with variances derived from rigorous error propagation can eliminate possible aliasing that might have occurred during sampling in the data acquisition stage (hardware filters). In addition, the filtered series contain much less data points thus making their analysis possible with small computer systems (e.g., a desktop computer).

2) The interpolation of gaps in the data series is unnecessary when using LSSA. This is efficient for the data analysis especially in the case when high-level noise (e.g., from earthquakes) exists in the series. A very simple way to avoid saturated data segments is to eliminate the said segment altogether and proceed with the analysis with no need to interpolate the data.

3) As shown, one of the advantages of the LSSA is that there is no need to smooth any spikes before analysis. However, the distinction between spikes and extended high level noise should be clearly identified because the latter will certainly distort the spectra and the information we are interested in.

4) A Parzen window is suggested for filtering (or decimating) the data series, but care must be exercised when selecting its bandwidth so as not to over-smooth the final spectrum (loss of essential information) or, at the other extreme, produce a very noisy spectrum in which the signal will not be detectable. For detecting the minute translational modes in the Earth's inner core from superconducting gravimeter records, further research is needed for the estimation of the optimum bandwidth of the data decimation filter. This is the subject of another investigation.

Acknowledgements

The principal author wishes to express her sincere thanks to the Chinese Scholarship Council for the support while at York University as a visiting scholar. In addition, the Canadian National Center of Excellence GEOIDE provided support for the project held by the S. Pagiatakis. The principal author is also grateful to the Department of Earth and Atmospheric Science of York University (now Dept. of Earth and Space Science and Engineering) as the host institution.

References

Crossley D, and J. Hinderer (1994), Global Geodynamics Project-GGP: Status Report 1994. http:///eas.slu.edu/GGP/ggpsr94.html.

Merriam, J. B., An Ephemeris for Gravity Tide Predictions at the Nanogal Level, *Geophys. J. Int.*, 108, 415-422, 1992.

Chen, YQ. (1988) Deformation Observation Data Process. Press of Surveying and Mapping. Beijing.

Pagiatakis S.D. (1999) Stochastic Significance of Peaks in the Least-squares Spectrum. *Journal of Geodesy*, 73: 67-78.

Pagiatakis S. D. (2000) Application of the Least-squares Spectral Analysis to Superconducting Gravimeter Data Treatment and Analysis Cahiers du Centre Européen de Géodynamique et Séismologie, 17:103-113.

Vanicek P (1969) Approximate Spectral Analysis by Least-squares Fit. *Astrophysics and Space Science*, 4:387-391.

Vanicek P (1971) Further Development and Properties of the Spectral Analysis by Least-squares. *Astrophysics and Space Science*, 12:10-33.

Wells D. E., Vanicek P., Pagiatakis, S. D. (1985) Least Squares Spectral Analysis Revisited. Tech Rep 84, Department of Surveying Engineering, University of New Brunswick, Fredericton.

Words and Expressions

deliberately [dɪˈlɪbərətli] *adv.* 故意地
dissertation [ˌdɪsəˈteɪʃn] *n.* (学位)论文；专题；论述；学术演讲
catalog [ˈkætəlɒg] *v.* 编目录
theoretical [θɪəˈrətɪkl] *adj.* 理论的
affiliation [əˌfɪlɪˈeɪʃn] *n.* 联系；从属关系
acknowledgment [əkˈnɒlɪdʒmənt] *n.* 承认；承认书；感谢
comparatively [kəmˈpærətɪvli] *adv.* 比较地；相当地
exhaustive [ɪgˈzɔːstɪv] *adj.* 无遗漏的；彻底的；详尽的；无遗的
treatise [ˈtriːtɪs] *n.* 论文；论述
gerund [ˈdʒerənd] *n.* 动名词
condensed [kənˈdensd] *v.* 浓缩
introductory [ˌɪntrəˈdʌktəri] *adj.* 介绍性的；入门的；初步的
arrangement [əˈreɪndʒmənt] *n.* 排列；安排
indebtedness [ɪnˈdetɪdnəs] *n.* 受恩惠；亏欠；债务
bibliography [ˌbɪbliˈɒgrəfi] *n.* (有关一个题目或一个人的)书目；参考书目
Harvard [ˈhɑːrvərd] 美国哈佛大学
alphabetically [ˌælfəˈbetɪkli] *adv.* 按字母顺序地

Unit 30　Translation Techniques for EST
——科技英语翻译技巧

一、科技英语的特点（English for Science and Technology）

1. 英语的特点：
(1) 英语是一种成分复杂的组合型语言；
(2) 英语为分析性语言；
(3) 英语的语言结构比较合理；
(4) 英语的词性转换较多；
(5) 英语的词汇组成异常复杂和丰富；
(6) 英语的发音与拼写脱节。

2. 科技英语的特点
(1) 以客观陈述为主，较多使用被动语句；
(2) 使用大量科技英语词语；
(3) 表达方式程序化；
(4) 大量使用非限定动词；
(5) 大量使用名词和介词短语；
(6) 复杂的长句较多；
(7) 条件句较多；
(8) 省略句较多；
(9) 采用的句子结构严谨合理；
(10) 动词的时态相对固定。

二、科技英语翻译的标准

翻译标准是指翻译活动中所必须遵循的准绳，是衡量译文质量的尺码，是译者致力追求的目标。长期以来，"信""达""雅"三个字被我国翻译工作者公认为翻译的标准。具体如下：

1. 明确（即"信"——准确）
它有两方面的含义：
(1) 确切：准确无误地表达原文的含义，忠实于原文。
(2) 明白：清楚明白地表达原文的意思，不能模糊不清，模棱两可。
例句：Few quantitative data are available.
译文：很少有定量数据可资利用（可用的定量数据很少）。

2. 通顺（即"达"——通顺、流畅）

指译文中应当符合中文的语法要求，易懂、读起来顺口。

(1) 选词造句正确，符合汉语习惯。

(2) 语气表达正确，恰当表达出原文的语气、情态、语态及强调的重点。

例句1　This possibility was supported to a limited extent in the tests.

译文1　在实验中，这一可能性在有限的程度上被支持了。

译文2　这一可能性在实验中于一定限度内得到了证实。

例句2　Influenza is spread in the same manner as a common cold.

译文1　流行性感冒以和普通感冒相同的方式被传播着。

译文2　流行性感冒的传播发生与普通感冒相同。

3. 简练（即"雅"——得体）

简练是指译文要尽可能简短、精练，没有冗词废字，不重复啰嗦。

例句1　The flow sheet shown in Fig. 2 is intended to illustrate the SL/RN process.

译文1　图2所示的流程图是想阐明SL/RN法的。

译文2　图2为阐明SL/RN法的流程。

例句2　Each product must be produced to rigid quality standard.

译文1　每件产品都必须生产符合严格的质量标准。

译文2　每件产品须达到严格的质量标准。

科技英语翻译最重要的是准确性和严密性，因此，在明确、通顺、简练这三条标准中，最重要的是明确，其次是通顺和简练。对于一般的科技资料的翻译，能做到明确和通顺就可以了，对于质量要求较高的翻译，则应当同时满足简练的要求。

三、翻译过程

翻译包括理解原文和表达译文这两个过程。

翻译过程：

(1) 理解原文——透彻了解原文的词义、语法关系和技术含义，以掌握原文的内容和实质。

(2) 表达译文——在忠实原文的基础上，运用一定的翻译技巧，写出流畅、简洁、通俗易懂的译文。

理解原文与表达译文两者关系是相互联系、不能截然分开。在理解过程中含有表达的因素，而在表达过程中又会促进理解和深化。

1. 理解过程

理解过程包括：(1) 辨明原文词义；(2) 辨明语法关系；(3) 辨明专业内容。

(1) 辨明原文词义——是理解原文的重要环节。

①联系上下文及专业范围来确定词义。

英语里一词多义的现象很普遍，词义往往因上下文、场合及专业范围而不同。因而要根据具体情况正确选定原句中的单词或词组的词义。

例句1　No signal transmitted from the cat and mouse station was detected.

译文　没有检测到从航向和指挥电台发出的信号。

例句2　The most precise method of determining elevations and most commonly used method is spirit leveling.

译文　测定高程精度最高和最常用的方法是几何水准测量。

②根据词类确定词义。

英语中一词多类的现象很普遍，类别不同，词义往往也不同。

例句1　Now the signal has input to the computer.

译文　现在信号已输入计算机。

例句2　Now that the signal has input to the computer, it can operate.

译文　既然信号已输入计算机，就可工作了。

例句3　The transmitter works well.

译文　发射机运作良好。

例句4　Works about microcomputers are particularly desirable at present.

译文　有关微型计算机的著作目前特别需要。

例句5　In general, the surveys necessary for the works of human beings are plane surveys.

译文　凡是为人工建筑物所进行的测量都是平面测量。

③仔细辨明词义和微细差别。

要准确理解词义，还需吃透词义，掌握其分寸，不能忽视用词之间的细微差别。

例句1　He is not a chemist.

译文　他不是化学家。

　　　　He is no chemist.

译文　他对化学一窍不通。

例句2　I had the radio repaired.

译文　我已请人修好这台收音机了。

　　　　I had repaired the radio before you came.

译文　你来之前，我已修好这台收音机了。

(2)辨明语法关系——理解原意很关键的一步。

辨明语法关系，包括判明句型结构，找出句子的各个成分，弄清各个成分的并列关系或从属关系。一般按下列步骤入手：

①通读全文。根据主语、谓语的数目及有无连词，确定句子是属于简单句、并列句还是主从复合句。

②找出各个句子的主要成分(主、谓、宾、表等)，进一步判明各次要成分(定、状、补、同位、插入等)与主要成分之间的关系。判明各个成分的涉及范围及其内部组成的层次。

③如果不是简单句，就要进一步判明各从句之间的并列关系或从属关系。

例句1　If we gather together all the basic equations which define the basic electrical units which we have discussed so far, we will find that if we know three such circuit parameters as voltage, current, and elapsed time, we can determine any of other parameters such as resistance, work and power.

译文　如果把迄今所讨论过的确定电的基本单位的所有基本公式都集中起来，我们就会发现：只要知道三个电路参数，电压、电流和经历时间，就能够求其他任一参数，如电阻、功和功率。

例句2　Numerous portable electronic devices making use of these wave phenomena have been developed which permit the measurement of distance with tremendous precision.

译文　利用波动现象的许多轻便型电子仪器现已研制出来，这些仪器使测距精度很高。

（3）辨明专业内容——是理解原文的重要步骤。

既然语法关系明白了，如果缺乏有关专业内容的理解，则还不能确切掌握原文的思想内容，译文将生硬别扭或偏离愿意。

例句1　Usually these conditions are determined at each end of a line and averaged. For long distance it may be desirable to determine the condition at intermediate points as well.

译文　通常在测线的两端测定气象条件，并取平均值，对于长距离测量还需中间点加测气象条件。（还应该将中间点的气象条件考虑进去）

2. 表达过程

表达过程分为：（1）理解原文是否透彻；（2）能否灵活运用各种翻译方法和技巧；逐字死译，则写不出流畅达意的译文，因此翻译方法与技巧，在科技英语中便显得十分重要。下面仅对科技英语中最常出现的被动语态的翻译作详细探讨，其他可参考有关文献。

四、科技英语中被动语态的翻译技巧

科技英语惯用被动语态，而汉语则很少采用明显的被动形式。虽然有时也采用"被"字来表示被动含义，但大多数场合下，这个"被"字无需写明就能反映被动含义。切忌照原文字面直译。

1. 译成主动句

（1）保留原主语与原句结构。

例句1　Sound waves have long been used for estimating distances.

译文　声波很久以来就用于估算距离。

例句2　As we have seen, a wave is transmitted from a transmitter to a reflector and returned.

译文　正如我们所看见的，波从发射器发射到反射器，然后再返回。

（2）主语移作宾语，而将原句中介词宾语适当译成主语。

例句1　Large quantities of lasers are used by optical communication in the generation of light emitting.

译文　光通信利用大量的激光器来产生光发射。

例句2　Distance is measured with EDM devices by determining the time required for an electromagnetic wave to travel from a transmitter at one end of line and to return from a reflector at the other end of line.

译文　电磁波测距仪通过测定从测线一端的发射器发射电磁波传到测线另一端的发射器上再返回到发射器所需时间来测定距离。

(3)主语移作宾语，译成无人称句。

例句1　These computers would be large and bulky, if vacuum tubes were used.

译文　要是采用真空管的话，这些计算机就会又大又笨。

(4)原文无主动者，译文加上适当主动者。

例句1　Very significant advance have been made in the past couple of decades that open the door to more rapid progress in surveying and mapping.

译文　在过去二十几年中，计算机技术已有很大发展，为测绘更迅速的发展打开了道路。

(5)用 it 代替主语从句的句子，译成无人称或不定人称句。

例句1　Let it be assumed that the reduced level of A is known to be 100m above a particular datum.

译文　假定 A 点的归化高程高出某基准面 100 米。

这类句型还有：

It is said that...　　据说，有人说……

It is stress that...　　人们强调说……

It can been seen that...　　可见……

It must keep in mind that...　　必须记住……

It will be noted that...　　你会注意到……

It should be pointed that...　　必须指出……

It is demonstrated that...　　已经证明……

2. 译成被动句

强调被动者或被动动作，或出于语气及修辞需要。

(1)用"被"字表达。

例句1　Radio waves are also known as radiant energy.

译文　无线电也被认为是辐射能。

例句2　About three-fourths of the surface of the Earth is covered with water.

译文　地球表面大约四分之三(的面积)被水覆盖着。

(2)用"受""由""待""让"等字表达。

例句1　Everything on or near the surface of the Earth is attracted.

译文　地球上或地球附近的一切物体都受地球吸引。

例句2　The magnetic field is produced by an electric current.

译文　磁场由电流产生。

(3)用"是……的""为……所""予以""加以"等词语表达。

例句1　The satellite must be carefully inspected before launching.

译文　卫星在发射之前必须仔细地予以检查。

例句2　Electricity has been known for thousands of years.

译文　电为人所知已有几千年了。

3. 倒装被动句的译法

有些英语被动句常把谓语部分中的过去分词提到句首位置，形成倒装结构。这种倒装被动句在译时无特殊处理的必要，一般可按照原文顺序译成汉语的"……的是……"形式。

例句1　Presented in the paper are new data on this subject.

译文　本文介绍的是有关这一课题的一些新数据。

例句2　To be particularly considered are the following factors.

译文　应特别加以考虑的是下列一些因素。

Part IV

Study Abroad Application Guide

Unit 31 International Universities/ Schools in Geomatics

For the benefit of our students in searching for the international Universities/Schools in Geomatics, relevant international academic units in Geomatics are summarized as follows in sequence of A to Z.

Australia
- Curtin University of Technology (CUT), Dept. of Spatial Sciences, Australia
  科廷(科技)大学,空间科学系
  网址:http://spatial.curtin.edu.au/
- U. of Melbourne, Dept. of Geomatics, Australia
  墨尔本大学,测绘系
  网址:http://www.sli.unimelb.edu.au/
- U. of New South Wales (UNSW), School of Surveying & Spatial Information Systems, Australia
  新南威尔士大学,测量与空间信息系统学院
  网址:http://www.gmat.unsw.edu.au/
- The Royal Melbourne Institute of Technology (RMIT), The School of Mathematical and Geospatial Sciences, Australia
  皇家墨尔本理工大学,数学与地球空间科学学院
  网址:http://www.rmit.edu.au/browse;ID=vpegh7ua2ata1
- U. of Southern Queensland (USQ), Faculty of Engineering and Surveying, Australia
  南昆士兰大学,工程与测量系
  网址:http://www.usq.edu.au/engsurv
- U. of Tasmania (UTAS), Surveying and Spatial Science, Australia
  塔斯马尼亚大学,测量与空间科学组
  网址:http://www.utas.edu.au/spatial/

Austria
- Graz University of Technology (GUT), Faculty of Technical Mathematics and Technical Physics, Austria
  格拉茨技术大学,科技数学与物理系
  网址:http://portal.tugraz.at/portal/page/portal/TU_Graz/Einrichtungen/Fakultaeten/FakMathematik

Brazil
- U. do Estado do Rio de Janeiro (UERJ), Departamento de Engenharia

Cartográfica, Brazil (in Portuguese)
里约热内卢州立大学，制图工程系
网址：http://www.carto.eng.uerj.br/

- U. Estadual Paulista (UNESP), Departamento de Cartografia, Brazil (in Portuguese)
保利斯塔州立大学，地图制图系
网址：http://www.fct.unesp.br/index.php?CodigoMenu=40&CodigoOpcao=40&Opcao=22

- U. Federal do Paraná (UFPR), Geomatics Department, Brazil
巴拉那联邦大学，测绘系
葡萄牙语网址：http://www.geomatica.ufpr.br/
英语网址：http://www.geomatica.ufpr.br/English.html

- U. Federal de Pernambuco, Departamento de Engenharia Cartográfica, Brazil (in Portuguese)
伯南布哥联邦大学，制图工程系
网址：http://www.ufpe.br/decart/

Canada

- U. of Calgary, Dept. of Geomatics Engineering, Canada
卡尔加里大学，测绘工程系
网址：http://www.geomatics.ucalgary.ca/

- U. of New Brunswick (UNB), Dept. of Geodesy and Geomatics Engineering, Canada
新不伦瑞克大学，大地测量与测绘工程系
网址：http://gge.unb.ca/HomePage.php

- York University, Department of Earth & Space Science & Engineering, Canada
约克大学，地球、空间科学与工程系
网址：http://science.yorku.ca/Schools-Departments/Earth-Space-Science-Engineering/

- Ryerson University, Department of Civil Engineering, Canada
瑞尔森大学，土木工程系
网址：http://www.ryerson.ca/civil/

Finland

- Aalto University, Department of Surveying, Finland
阿尔托大学理工学院(原赫尔辛基理工大学)，测量系
网址：http://maa.aalto.fi/en/

France

- École Nationale des Sciences Géographiques (ENSG), France
国立地理科学学院
网址：http://www.ensg.eu/

- Institut Géographique National (IGN), France

国家地理研究所
网址：http://www.ign.fr/

Germany
- U. Karlsruhe, Institute of Geodesy, Germany
  卡尔斯鲁厄大学，大地测量学院
  网址：http://www.gik.uni-karlsruhe.de/
- U. of Stuttgart, Institute of Photogrammetry, Germany
  斯图加特大学，摄影测量学院
  网址：http://www.ifp.uni-stuttgart.de/
- Ludwig-Maximilians-Universität München, Academic Sites for Geomatics Engineering, Germany
  慕尼黑大学，测绘工程学术网站
  网址：http://www.lrz-muenchen.de/%7et5831aa/WWW/Links.html
- Leibniz Universität Hannover, Faculty of Civil Engineering and Geodetic Science, Germany
  汉诺威大学，土木工程与大地测量学院
  网址：http://www.uni-hannover.de/en/fakultaeten/fk-bauge/index.php

Greece
- National Technical University of Athens, Faculty of Rural & Surveying Engineering, Greece
  雅典国立科技大学，农村与测量工程系
  网址：http://www.survey.ntua.gr/

Hong Kong
- The Hong Kong Polytechnic University (HKPU), Dept. of Land Surveying & Geo-Informatics, Hong Kong
  香港理工大学，土地测量与地理资讯学系
  网址：http://www.lsgi.polyu.edu.hk/

Ireland
- Dublin Institute of Technology (DIT), Department of Spatial Information Sciences, Ireland
  都柏林理工学院，空间信息科学系
  网址：http://www.dit.ie/spatialplanning/information-sciences/

Netherlands
- Delft University of Technology, MSc in Geomatics, Netherlands
  代尔夫特理工大学，测绘学硕士
  网址：http://geomatics.tudelft.nl/
- International Institute for Aerospace Survey and Earth Sciences (ITC), Netherlands
  国际航天测量与地球学学院
  网址：http://www.itc.nl/
- Utrecht University (UU), Department of Earth Sciences, Netherlands

乌得勒支大学，地球科学系

网址：http://www.uu.nl/faculty/geosciences/EN/facultystructure/departments/departmentearthsciences/Pages/default.aspx

New Zealand

- U. of Otago, School of Surveying, New Zealand
  奥塔哥大学，测量学院
  网址：http://www.surveying.otago.ac.nz/

Palestine

- Palestine Polytechnic University (PPU), Department of Civil and Architectural Engineering, Palestine
  巴勒斯坦理工大学，土木建筑工程系
  网址：http://cet.ppu.edu/index.php? option = com _ content&view = article&id = 263&Itemid = 318&lang = en

Poland

- U. of Warmia and Mazury in Olsztyn, Faculty of Geodesy and Land Management, Poland
  奥尔什丁瓦尔密亚玛祖里大学，大地测量与土地管理系
  波兰语网址：http://www.geo.kortowo.pl/
  英语网址：http://www.geo.kortowo.pl/english/

Portugal

- U. of Lisbon, Dept. of Geographic Engineering, Geophysics and Energy, Portugal
  里斯本大学，地理工程、地球物理与能源系
  网址：http://enggeografica.fc.ul.pt/inicio_UK.htm

Russia

- Moscow State University of Geodesy and Cartography, Russia (in Russian)
  莫斯科国立测绘大学
  网址：http://www.miigaik.ru/

South Africa

- U. of Cape Town (UCT), School of Architecture, Planning and Geomatics, South Africa
  开普敦大学，建筑、规划与测绘学院
  网址：http://www.geomatics.uct.ac.za/

Sri Lanka

- Sabaragamuwa University of Sri Lanka (SUSL), Faculty of Geomatics, Sri Lanka
  斯里兰卡萨伯勒格穆沃大学，测绘系
  网址：http://www.sab.ac.lk/geo/index.htm

Sweden

- Royal Institute of Technology (KTH), Department of Urban Planning and Environment, Sweden
  瑞典皇家理工学院，城市规划与环境系

网址：http://www.kth.se/abe/inst/som？l=en_UK
其下大地测量学分支：http://www.infra.kth.se/geo/indexeng.html

Turkey
- Istanbul Technical University (ITU), Geomatics Engineering Department, Turkey
  伊斯坦布尔科技大学，测绘工程系
  网址：http://geomatik.itu.edu.tr/
- Karadeniz Technical University, Department of Geodesy and Photogrammetry Engineering, Turkey (in Turkish)
  黑海科技大学，大地测量与摄影测量工程系
  网址：http://www.jeodezi.ktu.edu.tr/
- Yildiz Technical University, Department of Geodesy and Photogrammetry Engineering, Turkey
  伊尔迪兹科技大学，大地测量与摄影测量工程系
  网址：http://www.hrm.yildiz.edu.tr/web/english/index.php
  测绘相关链接：http://www.hrm.yildiz.edu.tr/web/english/Linkler.php？Bolum=7
- Zonguldak Karaelmas University, Geodesy and Photogrammetry Engineering Department, Turkey
  宗古尔达克大学，大地测量与摄影测量工程系
  网址：http://jeodezi.karaelmas.edu.tr/eng.htm

U.K.
- U. College London (UCL), Department of Civil, Environmental and Geomatic Engineering, U.K.
  伦敦大学学院，土木、环境与测绘工程系
  网址：http://www.engineering.ucl.ac.uk/blog/departments/civil-environmental-and-geomatic-engineering/
- U. of Glasgow, Department of Geography and Geomatics, U.K.
  格拉斯哥大学，地理与测绘系
  网址：http://www.geog.gla.ac.uk/
- U. of Newcastle upon Tyne, School of Civil Engineering and Geosciences, U.K.
  纽卡斯尔大学(新堡大学)，土木工程与地球科学学院
  网址：http://www.ceg.ncl.ac.uk/
- U. of Nottingham, Institute of Engineering Surveying and Space Geodesy, U.K.
  诺丁汉大学，工程测量与空间大地测量学院
  网址：http://www.nottingham.ac.uk/iessg

USA
- California State University (Fresno), Department of Civil and Geomatics Engineering, U.S.A.
  加州州立大学，土木与测绘工程系
  网址：http://www.csufresno.edu/geomatics
- Ferris State University, Surveying Engineering Department, U.S.A.

费里斯州立大学，测量工程系
网址：http://www.ferris.edu/cot/surveying/
- Massachusetts Institute of Technology（MIT），the Geodesy and Geodynamics group，U.S.A.
麻省理工大学，大地测量与地球动力学组
网址：http://www-gpsg.mit.edu/
- Ohio State University，Department of Civil，Environmental and Geodetic Engineering，U.S.A.
俄亥俄州立大学，土木、环境与大地测量工程系
网址：http://ceg.osu.edu/
- U. of Alaska Anchorage（UAA），Department of Geomatics，U.S.A.
阿拉斯加（安克雷奇）大学，测绘系
网址：http://www.uaa.alaska.edu/schoolofengineering/programs/geomatics/index.cfm
- U. of Florida，the Geomatics Program，U.S.A.
佛罗里达大学，测绘课程项目
网址：http://www.surv.ufl.edu/
- U. of Maine，Dept. of Spatial Information Science & Engineering，U.S.A.
缅因大学，空间信息科学与工程系
网址：http://spatial.umaine.edu/

Unit 32 Get Ready to Study Abroad

Find Your Interested Program

Every year, hundreds of Chinese students apply for study abroad programs and wish to have life-changing experiences. Whether you are interested in pursuing a higher degree to improve your knowledge and skills, or simply experiencing a cultural exchange that broadens your view and minds, there is always a program that will fit your needs.

There are numerous prestigious universities that offer internationally recognized programs and degrees, such as the Master of Engineering in Geomatics Engineering or Doctor of Philosophy in Geomatics Engineering, which may have similar degree titles but different course structures. If you are considering pursuing a postgraduate education abroad, this unit could serve as a valuable reference for your application.

Generally speaking, while the degrees and specializations offered in Geomatics Engineering may be similar between universities, the admission requirements and list of required documents can vary. Here we will use the University of Calgary in Canada as an example to highlight key information from their official website: https://schulich.ucalgary.ca/geomatics/programs/graduate. This summary aims to guide you on what to prepare and how to get ready for the application of Geomatics Engineering programs and specializations that suit your interests and career goals. Please check the relevant websites for more detail information.

1. **Degrees and Specializations Offered**

- Doctor of Philosophy (PhD), thesis-based
- Master of Science (MSc), thesis-based
- Master of Engineering (MEng), course-based. A course-based master's degree in Geomatics Engineering is usually considered a final degree.

Specializations:
- Positioning, Navigation and Wireless Location
- Geodesy, Remote Sensing and Earth Observation
- Digital Imaging Systems
- GIScience and Land Tenure
- Energy and Environmental Systems (Interdisciplinary Specializations)

- Environmental Engineering (Interdisciplinary Specializations)
- Wearable Technology WTEC (Interdisciplinary Specializations)

2. General Admission Procedures

Any student who wishes to become a registered student of the university must complete an online application for admission, pay the application fee, submit required documentation to demonstrate they meet the admission requirements, receive an offer of admission and pay the admission deposit, if required, before they are permitted to register in courses. (https://www.ucalgary.ca/pubs/calendar/grad/current/gs-a.html).

3. Application Deadline

Applicants should be clear for the application deadlines for the graduate programs. It's best to be as prepared early as possible. Application deadlines are available on the Future Students website: https://grad.ucalgary.ca/future-students/explore-programs. Here are the application deadlines for different degrees of Geomatics Engineering if you are an international student.
- PhD (thesis-based): Mar. 1 application deadline for admission on September 1; July. 1 for admission on January 1; Nov. 1 for admission on May 1; and Jan. 1 for admission on July 1.
- MSc (thesis-based): Same as above.
- MEng (course-based): March 1 application deadline for admission on September 1.

Before You Apply

Admission requirements are very important for international students. Therefore, befoce applying, ensure you check the admission information relevant to the graduate program of your interest from the website: https://grad.ucalgary.ca/future-students/international-students and find whether or not you meet the admission requirements set out by the Faculty of Graduate Studies and by your graduate program of interest.

1. Admission requirements

The Faculty of Graduate Studies sets out the minimum qualifications for admission. However, graduate program requirements may call for higher scores, additional documentation and testing. Meeting the minimum requirements does not guarantee entry into a graduate program as admission is competitive. Applicants must hold or obtain the following minimum qualifications to be admitted to the Faculty of Graduate Studies:
- Minimum 3.0 GPA on a 4.0 point system
- Minimum education: A four-year baccalaureate degree or its equivalent from a recognized institution is required for a master's program; a master's degree in

most cases is required for admission to a doctoral program. At least 4 years of study (Bachelor's) — Average of 80% for Master's program; Master's degree — Average of 80% for Doctoral program.

- **English language proficiency**: TOEFL—minimum score of 86 (Internet-based, with no section less than 20); IELTS—minimum score of 6.5 (with no section less than 6.0) (Academic version).

2. Graduate program requirements

Each graduate program recommends the admission of an applicant to the Faculty of Graduate Studies based on departmental considerations. For the degrees and specializations offered by the graduate programs of Geomatics Engineering, the admission requirements are shown as follows: (https://schulich.ucalgary.ca/geomatics/programs/graduate)

(1) **PhD (thesis-based)**:
- GPA: A minimum of 3.5 GPA on a 4.0 point system
- Minimum education: MSc degree or equivalent from a recognized institution
- Work samples: Successfully complete a written field of study examination and an oral examination on the thesis proposal.
- Documents: A CV, A statement of purpose
- Reference letters: Two
- Test scores: None

(2) **MSc (thesis-based)**:
- GPA: A minimum of 3.0 GPA on a 4.0 point system, over the past two years of full-time study (a minimum of 10 full-course equivalents or 60 units) of the undergraduate degree.
- Minimum education: A four-year baccalaureate degree, or equivalent from a recognized institution.
- Work samples: None
- Documents: A CV, A Statement of Purpose
- Reference letters: Two
- Test scores: None

(3) **MEng (course-based)**:
- GPA: A minimum of 3.0 GPA on a 4.0 point system, over the past two years of full-time study (a minimum of 10 full-course equivalents or 60 units) of the undergraduate degree.
- Minimum education: A four-year baccalaureate degree in geomatics, civil or electrical and computer engineering, or in physics, computer science, geography, or an equivalent degree, from a recognized institution.
- Work samples: None
- Documents: One official copy of your final transcript (and degree certificate if

degree is not clearly stated on transcript) in signed, sealed envelope from institution attended. Unofficial transcripts can be uploaded to the online application for initial evaluation purposes. Admission offers, however, are conditional upon receipt of official transcripts.
- Reference letters: None
- Test scores: None

When you are ready to apply

1. Application process

- Choose a graduate program of interest and check its admission requirements
- Find a potential graduate supervisor, if required by the graduate program
- Prepare your documentation
- Calculate your tuition and fees
- Create an eID and start your application online

Applications for admission to the Faculty should be submitted through the online application system at ucalgary.ca/future-students/graduate/apply.

2. Required documents

- Transcripts
- Study proposal
- Curriculum vitae
- Reference letters
- Standardized test score

It's important to note that required documents vary depending on the graduate program. Your program may not require all of the documents listed below or may require certain documents that are not listed here. Please check your program requirements to ensure you have the complete list of required documents.

1) **Official transcripts:** Official transcripts must be sent directly from all post-secondary institutions you have attended (even if you did not graduate) to the Faculty of Graduate Studies.

2) **Curriculum vitae or resume:** Some master's and doctoral programs require a current version of your curriculum vitae.

3) **Reference letters:** Contact your referees and ask if they would provide a letter in support of your application. It is customary to give them at least one month's notice.

4) **Study proposal or essay:** A study proposal provides a clear description of the research or study area that you would like to pursue. Many successful applicants take approximately one month to research and write the proposal.

5) **Writing sample**: Some graduate programs require a writing sample. A sample must be something you have written, independently in an undergraduate or graduate class.
6) **Standardized tests**: Check your program requirements to know which tests you will need to take and the minimum score required to be considered for admission. The time to schedule and successfully complete these tests varies, and so you will need to plan accordingly.
7) **Document translation**: If your original documents are not in English, you must obtain a notarized word-for-word English translation. Send the translated copy to your graduate program in addition to your official documents.
8) **Program documents**: After you submit your online application, you will receive an email with your UCID number and instructions on how to access your Student Centre, where you will find a to-do list for providing program-specific supporting documents.

Words and Expressions

specialization [ˌspeʃəlaɪˈzeɪʃn] n. 专业；专门化；特殊化
interdisciplinary specialization 交叉学科专业
tenure [ˈtenjər] n. 保有；保有条件；占有；任职；终身职位；任期
land tenure 土地使用权；土地所有制；土地占有制
wearable [ˈwerəbəl] adj. 可穿用的；可佩带的
wearable technology 可穿戴技术
baccalaureate [ˌbækəˈlɔːrɪət] n. 学士学位
bachelor [ˈbætʃələr] n. 学士；单身汉
transcript [ˈtrænskrɪpt] n. 学生成绩报告单；文本；转录物
curriculum vitae [ˌkʌrɪkjələm vaɪˈtaɪ] (CV) n. 个人履历
résumé (resume) [ˌrezjuːˈmeɪ] n. 简历
reference [ˈrefrəns] n. 引证；证明信；介绍信；引文
supervisor [ˈsuːpərvaɪzər] n. 研究生导师；监督者
customary [ˈkʌstəmeri] adj. 习俗的；习惯的；依照习惯的
notarize [ˈnoʊtəraɪz] v. 由公证人确认……合法；公证
notarized certificate 公证证明
equivalent [ɪˈkwɪvələnt] adj. 相等的；相同的；等同的 n. 对等的人；当量

Terms Highlights

degree and specialization 学位和专业
Doctor of Philosophy (PhD) 哲学博士
Master of Science (M.Sc.) 理学硕士

Master of Science in Engineering (M.Sc.E.) 工学硕士
Master of Engineering (M.Eng.) 工程硕士
general admission procedures 一般入学程序
admission requirements 入学要求
application deadline 申请截止日期
Faculty of Graduate Studies 研究生院
graduate program requirements 研究生专业要求
curriculum vitae 个人履历
résumé (resume) 简历
degree certificate 学位证书
academic credential 学历
personal statement 个人陈述
statement of purpose 目标陈述
official transcript 官方成绩单
study proposal 学习计划
research proposal 研究计划
reference letter 推荐信
writing sample 写作(样本)
standardized test 标准化考试
cover letter 求职信

Unit 33 How to Write a CV

Types of Personal Profiles

There are three different types of written documents related to personal records or profiles: curriculum vitae, resume, and biography. Depending on your objectives, whether it is pursuing higher education, seeking employment, or applying for an academic research position, you may need to choose the appropriate type of profile that best fits the requirements.

Curriculum vitae: A CV is a formal, detailed document highlighting a person's professional and academic history. The term is derived from the Latin phrase curriculum vitae meaning "course of life", where curriculum means course and vitae means life. A CV generally includes personal profiles (name, gender, citizenship, date of birth, place of birth, rank, link, email and phone as well as address), educational background, professional experience, technical expertise, publications and accomplishments. Due to the level of detail involved, a CV is typically several pages in length and is structured in a clear and organized manner. A CV is often used to apply for academic research and teaching positions in universities and research institutes.

Resume: A résumé, or resume, is a formal and succinct summary of an individual's education, work experiences, skills and qualifications that are relevant to the job or degree program being applied for. The term "resume" is derived from the French word résumé, meaning "summary". Typically, a resume is a one or two-page document that provides a brief yet comprehensive overview of an individual's education, professional experience, skills and qualifications. It is recommended to accompany the resume with a cover letter to provide additional explanation for why the applicant is a good fit for the role.

Biography: A bio, short for biography, is a concise written history of a person's life that provides a brief description of who they are and what they do. A bio is usually only a few lines long and includes the most important things a stranger should know about the person, such as the person's name, current occupation or role and most significant achievements.

Differences Between CV and Resume

A CV and a resume are similar in that they're both formal written documents that highlight a person's education, skills, professional history, and achievements. However, there are significant differences between the two.

(1) A CV is typically longer than a resume. For example, a CV is usually 2 to 4 pages for a young professional, 4 to 7 pages for a person with more experience, and should not exceed 10 pages. On the other hand, a resume is generally prepared with a maximum of two pages.

(2) A CV is a formal and chronologically detailed document focused on a person's professional research experience, academic background and publications and is typically used for academic research/teaching position application in a university or a research institute. In contrast, a resume is concise and tailored and highlights a person's education, work experiences, skills and qualifications, specifically relevant to the job or degree they are applying for.

(3) CVs and resumes are not interchangeable in the United States and most of European countries, while they may be interchangeable in Australia and some Asian or African countries. Therefore, you should take consideration of the cultural and regional norms of the industry and country you're applying to and choose the appropriate one to use.

How to Write a CV

A well-written CV can benefit you and give an employer's good impression for your professional and academic credentials. So it's important to structure it in a way that best showcases your accomplishments and experience. Follow these steps when writing a CV:

(1) Personal profile: Include your personal information such as name, gender, citizenship (if beneficial), date of birth, place of birth, rank, link, email and phone as well as address.

(2) Educational history: Include all information on Bachelor, Master, Ph. D., postdoctoral or other relevant studies, in reverse chronological order, i.e., most recent first, with degree awarded or anticipated, name of institution, location, date.

(3) Professional experience: List the name of company or organization, and location, job title and dates employed in reverse chronological order starting with your most recent job.

(4) Publications/Inventions: Include all publications and inventions in correct bibliographic format with relevant information such as author(s), title, publisher, and place of publication, name of periodical, volume, issue number, date, and page(s).

(5) Honors and awards: Include both academic and professional with the name of honor, granting institution or agency, and date.

(6) Professional affiliations: Include current memberships with dates.

An Example of CV

NAME	YIN Hui
GENDER	Female
RANK	Professor of Geomatics Engineering, School of Geodesy and Geomatics, Wuhan University
EMAIL	hyin@ sgg.whu.edu.cn
PHONE	027-68718888

EDUCATION

Ph.D	Geodesy and Geomatics Engineering. School of Geodesy and Geomatics, Wuhan University, China, 1998.
M.Sc.E.	Geodesy and Geomatics Engineering. Dept. of Engineering Surveying, Wuhan University, China, 1990.
B.Sc.E.	Engineering Surveying. Dept. of Engineering Surveying, Wuhan University, China, 1984.

EMPLOYMENT HISTORY

2000.07-2022.12	Professor at School of Geodesy and Geomatics, Wuhan University, China.
2002.10-2003.10	Visiting Professor and Research Assistant at Department of Earth and Space Science and Engineering, York University, Canada.
1996.12-2000.06	Associate Professor at School of Geodesy and Geomatics, Wuhan University, China.
1995.07-1996.07	Research Assistant at Department of Land Surveying and Geo-Informatics, Hong Kong Polytechnic University, Hong Kong.
1992.12-1996.11	Lecture at Dept. of Engineering Surveying, Wuhan University, China.
1984.07-1992.11	Assistant Lecture at Dept. of Engineering Surveying, Wuhan University, China.

UNDERGRADUATE/GRADUATE TEACHING

Undergraduate Courses
- Engineering Surveying (Bilingual Education)
- Principles of Geographic Information System
- Engineering Surveying

- Engineering Surveying Course Design and Practice
- Urban Disaster Emergency Response and Management
- Deformation Observation Data Process
- English for Geomatics Engineering
- Fortran Language and Application in Surveys

Graduate Courses
- Time Series Analysis and Its Enginfeering Application
- Deformation Analysis and Prediction Methodology
- System Analysis and Modeling
- Scientific Paper Writing in English

SELECTED PUBLICATIONS

Books:
- YIN Hui and WANG Jianguo. Engineering Surveying (in English). Wuhan: Wuhan University Press, 2022.
- YIN Hui, etc. English for Geomatics Engineering(2nd). Wuhan: Wuhan University Press, 2013.
- HUANG Shengxiang, YIN Hui & JIANG Zhen. Deformation Monitoring Data Process. Wuhan: Wuhan University Press, 2010.
- YIN Hui and Spiros Pagiatakis, etc. English-Chinese/Chinese-English Vocabulary Handbook for Geomatics Engineering. Wuhan: Wuhan University Press, 2008.
- YIN Hui. Theory and Methodology for Temporal-spatial Deformation Analysis and Forecasting. Beijing: Surveying and Mapping Press, 2002.

Journals
- Xiaoming Zhang and Hui Yin. A Monocular Vision-Based Framework for Power Cable Cross-Section Measurement[J]. Engergies, 2019, 12 (15): 3034(1-26).
- Wang J., and Yin H. Failure Rate Prediction Model of Substation Equipment Based on Weibull Distribution and Time Series Analysis[J]. IEEE Access, Vol. 7, 2019: 85297-85309.
- YIN Hui, ZHANG Xiaoming, LI Xiaoxiang, et al.. Galloping Information Extraction and Spectral Analysis for Transmission Lines Based on Video Monitoring[J]. High Voltage Engineering, 2017, 43(9): 2889-2895.
- YIN Hui, Zhou Xiaoqing, et al.. Non-equdistant GM (1, 1) Modeling Comparison and Application[J]. Engineering of Surveying and Mapping, 2017, 26(2): 1-4. (in Chinese).
- YIN Hui, Zhou Xiaoqing, et al.. Non-equdistant Multi-point Deformation Prediction Model and Its Application[J]. Acta Geodaetica et Cartogrphica Sinica, 2016, 45 (10): 1140-1147.
- YIN Hui, et al.. Full-English Teaching Mode Establishment and Implementation Strategies for Geomatics Engineering[J]. Engineering of Surveying and Mapping,

2015, 24(6):77-80.
- YIN Hui, WANG Yantao, ZHANG Xiaoming, et al.. A 3D Visualization Platform Design and Realization for Transmission Corridor Based on ArcEngine[J]. Journal of Geomatics, 2015, 40(2):14-16.
- YIN Hui, SUN Mengting, et al.. 3D Modelization of Transmission Tower Based on Sketch Up[J]. Bulletin of Surveying and Mapping, 2015(4):34-37.
- YIN Hui, ZHANG Xiaoming, WANG Yantao, et al.. Research on 3D Visualization of Electromagnetic Interference for Ultra-high Voltage Transmission Lines[J]. High Voltage Engineering, 2014, 40(12):3874-3881.
- YIN Hui. The Revelation to Our Geomatics Engineering Education from the Co-op Education in University of Waterloo[J]. China University Teaching, 2011(11):95-96.
- YIN Hui and WANG Yanyan. Detection of Weak Superconducting Gravimetric Signals with Product spectrum of Weighted LSSA[J]. Journal of Geodesy and Geodynamic, 2011,31(4):63-65.
- YIN Hui, YANG Rui and CHEN Pengyun. Stability Test Methodology of Monitoring Benchmarks and Its Application[J]. Journal of Wuhan University of Technology, 2010,32(6):861-862,872.

Words and Expressions

biography [baɪˈɒɡrəfi] n.传记；生平；传记文学；人生历程
succinct [səkˈsɪŋkt] adj.简洁的；简明的；简练的
chronological [ˌkrɒnəˈlɒdʒɪkl] adj.按时间顺序的；按先后顺序的；按时间计算的
tailor to the situation 根据情况量身定做
tailor [ˈteɪlər] vt.专门制作；订做
norm [nɔːrm] n.规范；准则
credential [krɪˈdenʃl] n.凭据；资质；证件；国书
showcase [ˈʃəʊkeɪs] v.展示；表现 n.玻璃陈列柜；展示场所
accomplishment [əˈkʌmplɪʃmənt] n.成就；成绩；造诣；技能
reverse-chronological 倒序的；按时间逆序排列的
reverse [rɪˈvɜːrs] adj.反向的；倒转的；颠倒顺序的
bibliographic [ˌbɪbliəˈɡræfɪk] adj.书目的；著书目录的；书籍解题的

Unit 34 Tips for Resume Writing

What Is Resume?

As described in Unit 33, a resume is a formal and concise document that summarizes a person's education, work experiences, skills and qualifications relevant to the job or degree program being applied for. The purpose of a resume is to showcase an individual's unique qualifications and experiences and persuade the employer to invite the person for an interview. Therefore, if you are ready to apply for a degree program or a job, it's important to have a strong and concise resume to grab the employer's attention and prompt them to invite you for an interview.

A resume is like an advertisement. It tells a potential employer who you are, what you have done, and what you can do. It also states what kind of work you are seeking. So it is very important to design your resume to match the position or degree you are applying for, and list relevant skills, knowledge and experience to attract the employer to evaluate your qualifications.

Major Sections of a Resume

Depending on the relevance of the job or degree you're applying for, it's important to determine which contents and sections to include in your resume. Please note you must showcase your skills, experience and accomplishments in an honest depiction. Here are major sections of a resume:

(1) **Contact Information**: Include your name, address, telephone number and email address as well as student identification number if you are a college student. This information is very important and should be present on both pages using a font size larger than the body.

(2) **Skills Summary**: Provide a concise summary of your qualifications, your relevant knowledge, skills and strengths to show your competitive advantage. Typically, four to six points can be outlined with nouns or adjectives.

(3) **Education**: Describe your education status (e.g. Bachelor, Master, Ph.D., postdoctoral or other relevant studies) in reverse chronological order, with degree awarded or anticipated, major or specialization, and university location and dates. You can also list three to six courses related to your application as a bullet point of the

Education section.

(4) **Work Experience**: Include up to six points that are most relevant to the job you are seeking in decreasing order of importance to outline your work experience.

(5) **Others**: Highlight strengths in these areas such as awards and scholarships, computer proficiency, professional memberships or certification, publications, volunteer experience as well as activities and interests.

Tips for Resume Writing

A well-written resume can attract a potential employer's attention and interest for your professional and academic career, but designing a favorable resume in an attractive style and well-organized appearance is a hard and time-consuming work. Here are useful tips for resume writing:

- Check the requirements from the job description before writing your resume, then create sections according to their relevance to the job and highlight your strengths. Generally, skill summary or summary of qualifications is the first section. Be sure to organize the contents such that your most important information is on the first page.
- Use key words listed in the job descriptions you are applying (e.g., "GNSS data analysis, computer programming, problem-solving") to increase the chance of your resume being selected during human or electronic scanning. The chronological style is the most common.
- Use bulleted statements throughout your resume. Each bullet point should be relevant to your qualifications and the position.
- Design the layout of sections, margins, font size and style. A well-designed layout can attract a potential employer's attention and interest by using balanced margins, font size and style. A popular font size is between 10 point to 14 point, and the common styles are Times New Roman, Arial, Palatino Linotype, etc.
- Send your resume together with your cover letter to the prospective employer in a common file type such as .doc or pdf. It's suggested to save the resume under your name (e.g. "jack li resume.pdf") so that the human resource manager can easily save down your profile.

Resume Samples

Sample one:
JACK L1   **WHU ID: 2019202148888**

129 Luoyu Road Wuhan China 430079   Phone: 12345678987   E-Mail: li129@163.com

SKILLS SUMMARY

- Strong mathematical foundation in Geomatics Engineering in quantitative analysis
- Proficient in C, C#, VBA, MATLAB, R, SQL, and MS Excel
- Highly motivated, hardworking, keen learner with great attention under pressure

EDUCATION

- Bachelor of Engineering (Candidate), Geomatics Engineering, Wuhan University, Hubei, China   9/2019—Present
- Graduate, No.1 Middle School Affiliated to Central China Normal University, Wuhan, Hubei, China   6/2019

WORK EXPERIENCE

Deformation Data Analyst — Model Validation, Spatial Information Technology Engineering Co., Ltd., Wuhan, Hubei   7/2022—Present
- Developed Excel — MATLAB automation tool to perform model selection based on time series analysis model
- Reverse engineered and documented coordinate framework for data management

Computer Programmer — Engineering Surveying, Technology Engineering Co., Ltd, Wuhan, Hubei   7/2021—8/2021
- Software developer based on C#
- Write user guide documents

AWARDS

- National Encouragement Scholarship for Outstanding Academic Performance in 2020—2021
- XIA Jianbai Innovation Scholarship for Outstanding Young in 2021—2022
- First-class Scholarship for Outstanding Academic Performance in 2019—2020
- First Prize for the National College Student Scientific Paper Competition in Geomatics in 2021

ACTIVITIES & INTERESTS

Geomatics Data Mining Cup Team Leader, Wuhan, Hubei   7/2021—8/2021
- Performed large data analysis on historical transactions of Geomatics Project
- Built statistical models to predict data behavior from historical data using regression

Member of Student's Union, Propaganda Department in Student's Union at School of Geodesy and Geomatics, Wuhan University, Wuhan, Hubei   9/2020—6/2021
- Collaborated with other students in designing posters for university-wide debate

competition, Alumni Forum
- Assisted in organizing the 2021 New Year Campus Carnival

Extracurricular hobbies: Doing sports, travelling, and painting.

Sample two:

Jack LI

Geomatics Engineering, School of Geodesy and Geomatics
Wuhan University ID: 2019202148888

Current Address		**Permanent Address**
129 Luoyu Road	li129@163.com	129 Luoyu Road
Wuhan, China, 430079	12345678987 (cell)	Wuhan, China, 430079

SKILLS SUMMARY
- Practical computer programming experience gained through post-secondary studies
- Basic knowledge of Scheme, C, Photoshop CS3, flash 8, iWork, Microsoft Office
- Fundamental Geomatics knowledge
- Analytical and communicative skills

EDUCATION
- Candidate for Bachelor of Engineering, Geomatics Engineering, Wuhan University, Hubei, China, 9/2019—Present
- Graduate, No.1 Middle School Affiliated to Central China Normal University, Wuhan, Hubei, China, 6/2019

WORK EXPERIENCE
- Quantitative Analyst — GNSS Data Analytics, Spatial Information Technology Engineering Co., Ltd., Wuhan, Hubei 7/2022—8/2022
- Computer Programmer — Engineering Surveying, Technology Engineering Co., Ltd., Wuhan, Hubei 7/2021—8/2021

AWARDS
- 2nd-class Scholarship, Wuhan University, Wuhan, Hubei, P.R.China, 9/2020
- Excellent Student Leader, Wuhan University, Wuhan, Hubei, P.R.China, 4/2021
- 2nd Prize in Military Training, Wuhan University, Wuhan, Hubei, P.R.China, 9/2019
- First Prize in the National College Student Scientific Paper Competition in Geomatics in 2021

ACTIVITIES & INTERESTS
- Member of Geomatics Engineering Student Association, Wuhan, Hubei 7/2021 — present
- Member of Student's Union, Propaganda Department in Student's Union at School of Geodesy and Geomatics, Wuhan University, Wuhan, Hubei, 9/2020—6/2021
- Extracurricular hobbies: Traveling, cooking, and photography

Words and Expressions

grab [græb] v.攫取；抓取；吸引……的注意力；使留下印象
prompt [prɒmpt] v.引起；激起；鼓励……往下说；追问；提示；促使……采取行动
interview [ˈɪntəvjuː] n.接见；会见；采访；面试 v.采访；询问；进行面试；审问；表现
depiction [dɪˈpɪkʃn] n.描写；描绘；刻画；叙述
font [fɒnt] n.字型；一副活字；洗礼盆；盛油罐
identification [aɪˌdentɪfɪˈkeɪʃn] n.辨认；识别；身份证明
student identification number 学号
noun [naʊn] n.名词
adjective [ˈædʒɪktɪv] n.形容词
membership [ˈmembərʃɪp] n.会员资格；会员身份
margin [ˈmɑːdʒɪn] n.页面的空白边；页边线；边；边沿；边缘
prospective [prəˈspektɪv] adj.预期的；盼望中的；未来的；未来可能发生的

Unit 35　Cover Letter Writing

What Is Cover Letter?

A cover letter is a formal document that accompanies your resume or CV when applying for a job with a potential employer or organization. It should be written in business format and is typically no longer than one page.

A good cover letter is usually customized to the specific position or situation and should highlight the most important information about your competitive advantage. It should only include job-relevant information, such as past work experience, educational background, qualifications, accomplishments, and skills related to the job. Therefore the information given in a cover letter is crucial and can determine the employer's first impression of you as a candidate.

Cover Letter Structure

How to write a cover letter and what should be included? Generally, a cover letter consists of several parts, including your contact information, a greeting or salutation, the letter's main body, and the closing, along with a signature.

(1) **Contact information**: Include your name, your mailing address, your home/cell number, your email address.

(2) **Greeting or salutation**: Address the recipient with a professional greeting or salutation. For example, Dear Recruiter.

(3) **Letter's main body**:
- First Paragraph: Name the job you are applying to and where you saw or heard advertisement; ensure that the subject line includes a job title and a job number or code, if one has been provided.
- Middle Paragraphs: Group your qualifications according to themes and prove points by using examples.
- Final Paragraph: Mention that you would like to provide more information in an interview.

(4) **Closing**: Close the cover letter in a professional and polite way. For example, Sincerely or Sincerely yours, Best regards or Regards, followed with a comma and a space.

(5) **Signature**: Complete name (first and last name), and title if applicable, along with your signature.

Letter Writing Tips

A well-written cover letter is an essential part of your job search which can attract a potential employer's attention and increase your chance of successes. It is no doubt that the best resume comes with a great cover letter attached. So writing a good cover letter is also a tough and time-consuming work. Here are useful tips for cover letter writing:

- Never send out a resume without a cover letter.
- Do not copy someone else's letters.
- Compose your cover letter to reflect your personality and carefully tailor to the situation.
- Do not duplicate your resume. Tailor different cover letters to different employers.
- Write in a business format with no longer than one page.
- Highlight the most important information to articulate your competitive advantage with only job-relevant information.
- Focus on your qualifications in writing a cover letter for resume.
- Sign the letter with your signature.

Basic Cover Letter Format

Your Name
Address
Phone
Email Address
Today's Date

Employer's Name
Company
Address

Dear (Recipient's Name),

First paragraph is to name the job you are applying to and where you saw or heard advertisement.

Middle paragraphs highlight the most important information which can articulate your competitive advantage such as your experience, education, accomplishments, and skills related to the job, as well as your personality.

> Final paragraph is to restate your interest in the company and like to provide more information if needed.
>
> Sincerely,
>
> Signature
> Title if applicable

Cover Letter Samples

129 Luoyu Road, Apt. 306
Hongshan District, Wuhan
Hubei, P.R. China, 430079
September 12, 2022

Dear Recruiter:

I am writing to you regarding your recent posted on https://psjobs-emploisfp.psc-cfp.gc.ca/..... for an APPLICATION of Spatial Data Analyst.

Attached is my current resume. I would like to summarize my strength as follows:

As a second-year postgraduate student at the School of Geodesy and Geomatics, Wuhan University, I have over two years of work experiences in related project research in Geomatics. I am proficient in Geospatial data process and analysis using mathematical and statistical techniques such as time series, modeling and numerical analysis for Geomatics Engineering. Extensive experiences include real property survey/drawing and area estimation based on AutoCAD, GPS/RTK technique to the engineering application as well as spatial information visualization and presentation by ArcGIS.

I have ability in problem solving, communication, research and project-management skills. My experience and skills make me a unique and a capable candidate to play a key role in Geomatics Engineering.

I would welcome the opportunity to further discuss my qualifications in relation to your needs.

Thank you for consideration, and I look forward to hearing from you soon.

Sincerely,

David Zhang
133466788987
zhang129@163.com

Jack LI	UW ID: *******
Phone: 519 *** ****	Geomatics
Email: li129@163.com	Faculty of Enviorment

Dear Officer:

I am writing to apply for a transfer to Co-op.

As a transfer student in the Faculty of Environment, I quickly overcame the barriers existing in culture and life, and adapted myself to the new environment here. In the orientation week, I passed UW English Language Proficiency Exam and met the First-Year English Writing Skills Requirement. During my first study term at University of Waterloo, I am diligent, hardworking and efficient in knowledge learning and applying, and have achieved good performance in the midterm examinations with an average score of 91.8 for five courses.

Moreover, I have one year's full-time academic study experience at School of Geodesy and Geomatics, Wuhan University, China, which enhanced my ability both in academic study and social work and have laid a solid foundation in math and computer science, which, I believe, will help me resolve the potential conflicts between study and work in the future. As for working experience, I served as a group leader of tutors for 340 students in my school, and designed posters and promotion materials for Propaganda Department in Student's Union.

Generally, I am analytical and efficient, and have abilities in time management, problem solving and communication skills. My experiences and skills make me a unique and capable candidate to transfer to Co-op.

Please find my resume enclosed. Thank you for consideration, and I look forward to hearing from you soon.

Yours sincerely,

Jack Li
Geomatics
Faculty of Environment
University of Waterloo

Words and Expressions

accompany [əˈkʌmpəni] v. 陪伴；护送；伴随；与……一起发生；为……伴奏；附加；兼带
customized [ˈkʌstəmaɪzd] adj. 定制的；自定义的
salutation [ˌsæljuˈteɪʃn] n. 致意；(信函中如 Dear Sir 之类的)称呼语
signature [ˈsɪɡnətʃər] n. 签名；署名
recipient [rɪˈsɪpiənt] n. 接受者；收件人
recruiter [rɪˈkruːtər] n. 招聘人员；征兵人员
sincerely [sɪnˈsɪəli] adv. 真诚地；诚实地
regards [rɪˈɡɑːdz] n. (用于信函结尾或转达问候)致意；问候
comma [ˈkɒmə] n. 逗号
tough [tʌf] adj. 艰苦的；艰难的；棘手的
duplicate [ˈdjuːplɪkeɪt] vt. 复制；复印；复写
articulate [ɑːˈtɪkjuleɪt] v. 流利连贯地表达；表达

Unit 36 Essays Related to University Application

Essay Definition and types

An essay is a short piece of writing that expresses a person's thoughts or opinions about a subject. Unlike poems or stories, essays present ideas and arguments in a non-fictional way, aiming to inform, persuade, or entertain the reader.

Essays come in many shapes and sizes and can cover a wide range of topics, including literature, art, and academics. Essays can also have different types, each with its own unique characteristics and purposes. These include narration, description, exposition and argumentation.

A narrative essay is to tell a story about your own life or experience. For instance, you might be asked to write a narrative essay to introduce yourself or express something about your personal qualities when applying for a university.

A descriptive essay uses vivid language and sensory details to describe a person, place or object you have seen or known well to create a lasting impression on the reader. For instance, you might be asked to write an essay to describe a person who has a strong personal influence on you.

An expository essay explains or describes a topic in an objective and concise manner. For instance, you might be asked to write an essay to describe the classification of GNSS positioning techniques and their characteristics.

An argumentative essay intends to persuade readers by presenting arguments or evidences on an issue. For instance, you might be asked to write an essay to make an evidence-based argument and develop a specific position on a topic of active learning and passive learning.

This unit mainly focuses on academic essays, which are written to share personal experiences or to present purely academic descriptions or expositions of a topic in a subjective writing form.

Essay, paper, thesis and dissertation are four popular terms that our university students often encounter in academic settings. What are the differences between these four terms? Papers, as described in Unit 29, focus on presenting professional views and research findings on specific topics, and can be published in journals or presented on academic conferences. Report paper and term paper are two types of documents frequently assigned in universities and colleges. Thesis and dissertation mainly refer to lengthy pieces of writing on a particular subject, often written to obtain a university degree. In contrast, essays are shorter in length and simpler in structure. They can range from as little as 500 words to as many as 5,000 words or more, but typically fall between 1,000 to 3,000 words. The most common structure for an essay includes an introduction, body paragraphs in support of statement, and a conclusion.

Essays Related to University Application

Universities may require different documents in their graduate school applications, but there are some documents that are commonly requested. These include the personal statement, statement of purpose, study proposal and research proposal which can typically be submitted through the university's application website. Some universities may even use the terms essay A or essay B to require the applicants to submit subjective writing pieces on different topics. Additionally, universities may request a writing, like a personal statement, that you have written yourself. All of these documents are considered as essays.

A personal statement is an integral component of university applications to showcase your achievements, qualifications, talents, and goals. It contains detailed information about your academic qualifications, related job experience, and achievements to demonstrate your personality, values, and motivation.

A statement of purpose is a concise essay that explains to the program's committee your reasons for pursing a graduate degree. It typically includes information about your background, qualifications, and achievements, as well as an explanation of why you are a great match for the program.

A study proposal or research proposal is a formal, structured essay that outlines the area of your interest, field of study, and specific focus of the intended project. It describes in detail what you plan to study, including the rationale for why it is worth researching, and how you intend to proceed with the project. A well-written proposal should include the major you want to study or the field of your research interest, an overview of the methodology, and a clear explanation of how you intend to approach the project.

The requirements for study proposals can vary across different graduate programs, with differences often seen in the length of the proposal. Additionally, there will be differences in the emphasis of the content based on the specific professional characteristics. The purpose of the study proposal is to convince the graduate committee or university that your research field and research goal are a good fit for the requirements of the degree program and manageable within the given the time and resource constraints.

From the website of the University of Calgary (https://grad.ucalgary.ca/future-students), the Faculty of Graduate Studies provides a very clear statement to guide you how to write a study proposal. Here we introduce the statement highlights from the Faculty of Graduate Studies directly as follows:

"Some graduate programs, especially thesis-based programs, will require you to submit a project study proposal. These proposals will vary in length and content depending on which program you choose.

When drafting your proposal, you should find an intersection of your own educational/ work/research experience and what the program offers. Typically, each department/program will have its own set of strengths and focused areas of expertise. You should be aware of these strengths and specialties and try to work these in to your own application. As a graduate student, you will be an integral (and important!) part of department life, so make sure to discuss how your program of work fits in to the broader research/teaching environment.

Course-based programs require statements that are more specific to the work involved with your degree program. Some of these statements involve details on your area of specialization, educational goals, and career expectations. Furthermore, professional course-based graduate degrees like a Master of Social Work or a Master of Education require relevant work and/or teaching experience that you will describe in your proposal application".

An Example of a Statement of Purpose

My long-term goal is to become a geomatics engineering researcher working in a government or research institute. Having realized that such a challenging path would require solid quantitative skills and comprehensive understandings of theory and methodology in geomatics, I find a graduate education in geomatics engineering vital for my knowledge and career development.

As a multidisciplinary student in Geomatics Engineering and Statistics, I have taken many excellent courses in mathematics, statistics, geomatics, and computer science during my undergraduate study at School of Geodesy and Geomatics, Wuhan University. These courses have provided me with a strong theoretical background in advanced geodesy and Earth observation topics and allowed me to analyze the global changes from both qualitative and quantitative perspectives. In addition to the regular coursework, the various small projects required by the curriculum have improved my ability to conduct in-depth research and solve complicated problems. In one of my projects, I performed extensive research and experiment before arriving at an appropriate nonparametric model to assess the impact of weather conditions on the crustal deformation. One thing I learned from this project is the importance of reflective thinking, innovation, and the ability to transcend individual disciplines. Moreover, the experience of taking extra course load has trained me to work efficiently, manage time effectively, and multitask under pressure.

While many geomatical topics have been covered in classes, it was during the work periods in Spatial Information Technology Engineering Co., Ltd. that I gained real-world exposure in quantitative geodetic data analysis. I was privileged to work on several projects where I developed automation tools using MATLAB and VBA to perform model selection based on time series analysis, and I also reverse engineered and documented coordinates framework for data management. This intensive programming experience has improved my knowledge in geodetic observation data process automation, software development, and algorithm optimization.

The Geomatics Engineering program's reputation in the University of Calgary as a leading geomatics program is well-known, but what makes this program unique is the breadth and depth of its curriculum, its notable faculty, and the pioneering integration of expertise in geomatics, statistics, and computer science. With an interdisciplinary undergraduate study in all these areas and working experiences using quantitative methods, I am confident that I have the prerequisite knowledge and technical skills needed for the demanding thesis-based MSc degree in the specialization of geodesy and Earth observation program. Moreover, my academic and professional experiences have instilled in me the right mindset to tackle future challenges and achieve my long-term career goals.

Words and Expressions

essay [ˈeseɪ] n. 论说文；散文；随笔；小品文
non-fictional [ˌnɒn ˈfɪkʃnəl] adj. 非虚构叙的
persuade [pəˈsweɪd] v. 说服；劝服；敦促；使相信；使信服；驱使
entertain [ˌentəˈteɪn] v. 给……娱乐；使快乐；对（意见；建议；感情）给予考虑
narration [nəˈreɪʃn] n. 叙述；故事；讲述；记叙文
narrative essay 记叙文
description [dɪˈskrɪpʃn] n. 描述；描写；描写文
descriptive essay 描写文
exposition [ˌekspəˈzɪʃn] n. 阐述；展览会；说明文
expository essay 说明文
argumentation [ˌɑːɡjumənˈteɪʃn] n. 推论；论证；议论文
argumentative essay 议论文
vivid [ˈvɪvɪd] adj. 生动的；逼真的；活跃的；充满活力的
sensory [ˈsensəri] adj. 感觉的；感官的
evidence [ˈevɪdəns] n. 根据；证明；迹象；痕迹；证据 v. 证实；证明
subjective [səbˈdʒektɪv] n. 对象；主题；问题；主语 adj. 主观的
encounter [ɪnˈkaʊntər] v. 遇到；面临
proposal [prəˈpəʊzəl] n. 提议；建议；建议书
integral [ˈɪntɪɡrəl] adj. 必需的；必要的；基本的；基础的
motivation [ˌməʊtɪˈveɪʃn] n. 动机；动力；诱因

constraint [kənˈstreɪnts] *n.* 限制；约束；束缚；局限；抑制；拘束；压抑；克制
nonparametric [ˌnɒnpærəˈmetrɪk] *adj.* 非参数的
transcend [trænˈsend] *v.* 超出；超越；胜过；超过
privilege [ˈprɪvəlɪdʒ] *n.* 特权；特惠待遇；优待；特别待遇
notable [ˈnəʊtəbl] *adj.* 值得注意的；显著的
instill [ɪnˈstɪl] *vt.* 逐渐灌输；徐徐滴入
mindset [ˈmaɪndset] *n.* 心理定势；心态
tackle [ˈtækl] *v.* 决心处理；与……交涉；与……磋商

Appendix I Vocabulary

abnormal [æb'nɔːməl] *adj.* 反常的；变态的
accompany [ə'kʌmpəni] *v.* 陪伴；护送；伴随；与……一起发生；为……伴奏；附加；兼带
accomplishment [ə'kʌmpliʃmənt] *n.* 成就；成绩；造诣；技能
accredited [ə'kreditid] *adj.* 可接受的；可信任的；公认的；质量合格的
acknowledgment [ək'nɔlidʒmənt] *n.* 承认；承认书；感谢
adjective ['ædʒiktiv] *n.* 形容词
adjustment [ə'dʒʌstmənt] *n.* 平差；调整；调节
adverse ['ædvəːs] *adj.* 不利的；敌对的；相反的
aerial survey 航测；航空测量；航空勘测
aerodynamic [ˌɛərəudai'næmik] *adj.* 空气动力学的
aerosol ['ɛərəsɔl] *n.* [物化]气溶胶；气雾剂；雾化器；浮质
affiliation [əˌfili'eiʃən] *n.* 联系；从属关系
airborne ['ɛəbɔːn] *adj.* 空气传播的；空降的；空运的
algebra ['ældʒibrə] *n.* 代数学
algorithm ['ælgəriðm] *n.* [数]运算法则
alidade ['ælideid] *n.* [测]照准仪
align [ə'lain] *v.* 使排列成行；排列
alignment [ə'lainmənt] *n.* 排成直线；路线
allowance [ə'lauəns] *n.* 容许误差；容差；容许量
alloy ['ælɔi] *n.* 合金
almanac ['ɔlməˌnæk] *n.* 年历；历书；年鉴
alphabetically [ˌælfə'betikli] *adv.* 按字母顺序地
alter ['ɔːltə] *v.* 改变
alternating current 交流电
altitude ['æltitjuːd] *n.* 高度；海拔；高地
altimetry [æl'timitri] *n.* 测高学，高度测量法（以海平面为基准）
ambiguity [ˌæmbi'gjuːəti] *n.* 意义不明确；含糊
analog ['ænəlɔg] *n.* 类似物；相似体
analogous [ə'næləgəs] *adj.* 类似的；相似的；可比拟的
analytical [ˌænə'litikəl] *adj.* 解析的；分析的
analytical geometry 解析几何
anchor ['æŋkə] *n.* 锚；顶梁柱；刹车

ancillary [æn'siləri] *adj.* 补助的；副的
anomaly [ə'nɔməli] *n.* 不规则；异常的人或物
antenna [æn'tenə] *n.* 天线
anticipated [æn'tisipeitid] *adj.* 预先的；预期的
antipodal [æn'tipədl] *adj.* 对跖的；正相反的
apex ['eipeks] *n.* 顶点
apparatus [ˌæpə'reitəs] *n.* 仪器；器械；设备
apportion [ə'pɔːʃən] *v.* 分配
appreciate [ə'priːʃieit] *vi.* 增值；涨价 *vt.* 赏识；鉴赏；感激
approximation [əˌprɔksi'meiʃən] *n.* [数]近似值；接近；走近
arbitrary ['ɑːbitrəri] *adj.* 任意的；武断的；独裁的；专断的
architect ['ɑːkitekt] *n.* 建筑师
argumentation [ˌɑːgjumən'teiʃn] *n.* 推论；论证；议论文
    argumentative essay 议论文
Aristotle ['æristɔtl] *n.* 亚里士多德(古希腊大哲学家、科学家)
arithmetic [ə'riθmətik] *n.* 算术；算法
arrangement [ə'reindʒmənt] *n.* 排列；安排
articulate [ɑː'tikjuleit] *v.* 流利连贯地表达；表达
artificial [ˌɑːti'fiʃl] *adj.* 人造的；假的；非原产地的
ascension [ə'senʃn] *n.* 上升
    right ascension 赤经
ascend [ə'send] *v.* 登上；攀登；升起；上升
    ascending node 升交点
assembly [ə'sembli] *n.* 集合；装配；集会；集结；汇编
assessment [ə'sesmənt] *n.* 估价；被估定的金额
association [əˌsəusi'eiʃən] *n.* 协会；联合；结交；联想
astrogeodesy ['æstrəudʒiəudisi] *n.* 天文大地测量学；天体测量说
astrogeodetic [ˌæstrəudʒiːəud'tik] *adj.* 天文大地(测量)的
Atlantic Ocean 大西洋
atmospheric [ˌætməs'ferik] *adj.* 大气的
attitude ['ætiˌtjuːd] *n.* 姿态；姿势；态度；看法
attribute [ə'tribjuː)t] *n.* 属性；品质；特征；加于；归结于
attune [ə'tjuːn] *vt.* 使相合；使合调
augment [ɔːg'ment] *v.* 增加；增大
autonomous [ɔː'tɔnəməs] *adj.* 自治的；独立自主的；自律的；自制的
aviation [ˌeivi'eiʃən] *n.* 航空；航空工业；飞行(术)
awareness [ə'wɛənis] *n.* 知道；晓得
azimuth ['æziməθ] *n.* 方位；方位角
    azimuthal projection 方位投影
baccalaureate [ˌbækə'lɔriət] *n.* 学士学位

bachelor ['bætʃələ] n. 学士；单身汉

backsight ['bæksait] n. 后视

ballistic [bə'listik] adj. 弹道的；弹道学的

barometer [bə'rɔmitə] n. 气压计

bearing ['bɛəriŋ] n. 方向；方位

bent [bent] n. 倾向；歪

Bessel Correction 贝塞尔改正

bibliographic [ˌbibliə'græfik] adj. 书目的；著书目录的；书籍解题的

bibliography [ˌbibli'ɔgrəfi] n. （有关一个题目或一个人的）书目；参考书目

biography [bai'ɔːgrəfi] n. 传记；生平；传记文学；人生历程

bisect [bai'sekt] v. 切成两分，对（截）开

BIW (body in white) 白车身（完成焊接但未涂装之前的车身）

blend [blend] n. 混合

block [blɔk] v. 阻塞；阻碍；堵塞 n. 块；块体；区；障碍

blunder ['blʌndə] n. 粗差；大错；失误

blur [bləː] v. 把（界线、视线等）弄得模糊不清；涂污；污损（名誉）；弄污

branch [brɑːntʃ] n. 分部；分店；（学科）分科；部门；支流；支脉

buffering ['bʌfəriŋ] n. 缓冲（作用）；减震；阻尼；隔离

bulky ['bʌlki] adj. 大的；容量大的；体积大的

bull's-eye ['bulzai] n. 靶心

bundle ['bʌndəl] n. 束；捆；包；一套；一批

bylaw ['bailɔː] n. 次要法规；（社、团制定的）规章制度

cadastre [kə'dæstə] n. 地籍簿；地籍；地籍图

    cadastral surveying 地籍测量

calendar ['kælində] n. 日历；历法

calibrate ['kælibreit] v. 校准

calibration [ˌkæli'breiʃən] n. 校准；标度；刻度

campaign [kæm'pein] n. ［军］战役；（政治或商业性）活动；竞选运动

canopy ['kænəpi] n. 树冠层；天篷；顶篷 v. 用顶篷遮盖

carrier ['kæriə] n. 搬运者；载体

Cartesian [kɑː'tiːziən] adj. 笛卡儿的；笛卡儿哲学的 n. 笛卡儿坐标系

cartographic [ˌkɑːtə'græfik] adj. 制图的；地图的

cast iron [ˌkæst'aiən] 铸铁；生铁

catalog ['kætəlɔg] v. 编目录

catchment ['kætʃmənt] n. 集水；集水处，汇水

category ['kætigəri] n. 种类；别；［逻］范畴

celestial [si'lestjəl, si'lestʃəl] adj. 天上的

    celestial sphere 天球

census ['sensəs] n. 人口普查；（官方的）统计

centrifugal [sen'trifjugəl] adj. 离心的

certification [ˌsəːtifiˈkeiʃən] n. 证明
charge [tʃɑːdʒ] n. 电荷；负荷；费用；主管；掌管；充电；充气；装料
chart [tʃɑːt] n. 图表；海图
chord [kɔːd] n. 弦；弦长
chronological [ˌkrɔnəˈlɔːdʒikl] adj. 按时间顺序的；按先后顺序的；按时间计算的
circumference [səˈkʌmfərəns] n. 圆周；周围
clamp [klæmp] n. 夹子；夹具；夹钳
clinometer [klaiˈnɔmitə] n. 测角器；倾斜仪
clockwise [ˈklɔkwaiz] adj. 顺时针方向的
cluster [ˈklʌstə] vi. 丛生；成群
CNC abbr. [计] Computerized Numerical Control 电脑数值控制
coarse [kɔːs] adj. 粗的；粗糙的
coding [ˈkəudiŋ] n. 译码
coefficient [ˌkəuiˈfiʃənt] n. [数]系数
coherent [kəuˈhiərənt] adj. 一致的；连贯的
coin [kɔin] n. 硬币；金属货币 v. 创造（新词语）；首次使用；铸币；制造
collaboration [kəˌlæbəˈreiʃən] n. 协作；通敌
collective [kəˈlektiv] adj. 聚集而成的；集体的；共同的；集合的
comma [ˈkɔːmə] n. 逗号
commence [kəˈmens] v. 开始；着手
commission [kəˈmiʃən] n. 委员会
commonwealth [ˈkɔmənwelθ] n. 国民整体；共和国；联邦
community [kəˈmjuːnəti] n. 团体；社会；（政治）共同体
comparatively [kəmˈpærətivli] adv. 比较地；相当地
compass [ˈkʌmpəs] n. 罗盘；指南针；圆规
compensator [ˈkɔmpenseitə] n. 补偿器
complexity [kəmˈpleksiti] n. 复杂(性)；复杂的事物；复杂性
composite [ˈkɔmpəzit] adj. 合成的；混合的；复合的 n. 合成物；混合物；复合材料
comprehensive [ˌkɔmpriˈhensiv] adj. 全面的；广泛的；能充分理解的；包容的
compromise [ˈkɔmprəmaiz] n. 妥协；折中
concentric [kɔnˈsentrik] adj. 同中心的
conceptional [kənˈsepʃənl] adj. 概念上的
Concise Oxford Dictionary 简明牛津字典
condensed [kənˈdenst] v. 浓缩
cone [kəun] n. [数、物]锥形物；圆锥体；（松树的）球果
conform [kənˈfɔːm] vt. 使一致；使遵守；使顺从
conformal [kənˈfɔːml] adj. [数]（地图投影中）正形投影的；等角的
  conformal map 保角映像；保形变换图
conformality [kənfɔːˈmæliti] n. 正形；正形性
conformance [kɔnˈfɔːməns] n. 顺应；一致

conformity [kən'fɔːmiti] *n.* 一致；符合
confuse [kən'fjuːz] *vt.* 使糊涂；搞乱
congress ['kɔŋgres] *n.* （代表）大会；[C~]（美国及其他国家的）国会；议会
conical ['kɔnikəl] *adj.* 圆锥的；圆锥形的
    conic projection 圆锥投影
    conics ['kɔniks] *n.* 圆锥曲线论；锥线论
conscientious [ˌkɔnʃi'enʃəs] *adj.* 尽责的
consequence ['kɔnsikwəns] *n.* 结果；[逻]推理；推论；因果关系；重要的地位
consistency [kən'sistənsi] *n.* 一致性；连贯性；连接；结合；坚固性
conspicuous [kən'spikjuəs] *adj.* 显著的
constellation [ˌkɔnstə'leiʃən] *n.* [天]星座；星群
constituent [kən'stitjuənt] *adj.* 有选举权的；有宪法制定[修改]权的；组成的
constraint [kən'streint] *n.* 限制；约束；束缚；局限；抑制；拘束；压抑；克制
consultant [kən'sʌltənt] *n.* 顾问；商议者；咨询者
consume [kən'sjuːm] *vt.* 消耗；消费；消灭；吸引
contemporary [kən'tempərəri] *n.* 同时代的人
converge [kən'vɜːdʒ] *v.* 汇聚；集中；（向某一点）相交；[数]收敛
conversion [kən'vəːʃən] *n.* 变换；转化
corresponding [ˌkɔri'spɔndiŋ] *adj.* 对应的；相关的；符合的；一致的
cost-effective [ˌkɔːst-i'fektiv] *adj.* 费用低廉的
council ['kaunsil] *n.* 理事会；委员会；参议会；讨论会议；顾问班子；立法班子
counterclockwise [ˌkauntə'klɔkwaiz] *adj.* 反时针方向的
counterpart ['kauntəpɑːt] *n.* 副本；极相似的人或物；配对物
crane [krein] *n.* 起重机；吊车；鹤 *v.* 吊运；伸长（脖子）看
credential [krə'denʃl] *n.* 凭据；资质；证件；国书
crew [kruː] *n.* 全体人员；（工作）队
criteria [krai'tiəriə] *n.* 标准
crumbling ['krʌmbliŋ] *adj.* 破碎的；摇摇欲坠的
crust [krʌst] *n.* 外壳；硬壳；面包皮
crystal ['kristl] *adj.* 结晶状的 *n.* 水晶；水晶饰品；结晶；晶体
    liquid crystal displays (LCDs) 液晶显示
curb [kəːb] *n.* 路边
curiosity [ˌkjuəri'ɔsəti] *n.* 好奇心
current ['kʌrənt] *n.* 电流；水流；气流；涌流；趋势
curriculum vitae 个人履历
curvature ['kəːvətʃə] *n.* 曲率；弯曲
customary ['kʌstəməri] *adj.* 习俗的；习惯的；依照习惯的
customized ['kʌstəmaizd] *adj.* 定制的；自定义的
cylinder ['silində] *n.* 圆筒；圆柱体；汽缸；柱面
cylindrical [sə'lindrikl] [计]圆柱的

cylindrical projection 圆柱投影
dam [dæm] n. 水坝；障碍
decay constant 衰变常数；裂变常数
decode [ˌdiːˈkəud] v. 译；解读；转换
decrease [diˈkriːs] v. 减少
dedicate [ˈdedikeit] vt. 献(身)；致力；题献(一部著作给某人)
deduce [diˈdjuːs] vt. 推论；演绎出
defect [diˈfekt] n. 过失；缺点
deformation [ˌdiːfɔːˈmeiʃən] n. 变形
deliberately [diˈlibərətli] adv. 故意地
demise [diˈmaiz] n. 死亡；让位；禅让 vt. 让渡；遗赠；转让
departure [diˈpɑːtʃə] n. 横距；偏移
depict [diˈpikt] vt. 描述；描写
depiction [diˈpikʃn] n. 描写；描绘；刻画；叙述
deploy [diˈplɔi] v. 部署；利用
depress [diˈpres] vt. 压下；压低；使沮丧
derivative [diˈrivətiv] n. 导数；微商
description [diˈskripʃn] n. 描述；描写；描写文
  descriptive essay 描写文
descriptor [diˈskriptə] n. [计] 描述符；描述子
detonate [ˈdetəneit] v. 爆炸；引爆
deviation [ˌdiːviˈeiʃən] n. 偏差；偏移
dexterity [deksˈterəti] n. 灵巧；机敏
diameter [daiˈæmitə] n. 直径
differential [ˌdifəˈrenʃl] adj. 差分的；差动的；微分的
differentiate [ˌdifəˈrenʃieit] v. 区别；区分
diffusion coefficient 扩散率；扩散系数；漫射系数
digitize [ˈdidʒitaiz] v. [计]将资料数字化
digitizer [ˈdidʒitaizə] n. 数字转换器
dimensional [diˈmenʃənəl] adj. 空间的
diminish [diˈminiʃ] v. (使)减少；(使)变小
diode [ˈdaiəud] n. 二极管
  light emitting diode displays (LEDs) 发光二极管显示
disaster [diˈzɑːstə] n. 灾难；天灾；灾祸
discrepancy [disˈkrepənsi] n. 差异；相差；矛盾
discrimination [disˌkrimiˈneiʃən] n. 辨别；区别；识别力；辨别力；歧视
disparity [diˈspæriti] n. 明显差异；悬殊；视差
dispersion [diˈspəːʃən] n. [数] 离差；差量；散布；驱散；传播；散射
dissemination [diˌsemiˈneiʃən] n. 分发
dissertation [ˌdisə(ː)ˈteiʃən] n. (学位)论文；专题；论述；学术演讲

distinguish [di'stiŋgwiʃ] v. 区别；辨别
distortion [di'stɔːʃən] n. 变形；扭转；歪扭；歪曲作用；反常；畸变
distribution [ˌdistri'bjuːʃən] n. 分配；分发；配给物；区分；分类
diversity [dai'vəːsiti] n. 多样性；差异
divisor [di'vaizə] n. 除数；约数
domain [dəu'mein] n. 领域；领土；领地；（活动、学问等的）范围
domestic [də'mestik] adj. 国内的；家庭的
Doppler ['dɔplə] adj. ［亦作 d-］（奥地利物理学家）多普勒的
drain [drein] n. 排水沟；消耗；排水
drone [drəun] n. （遥控的）无人驾驶飞机（或导弹）；雄蜂；嗡嗡声 v. 嗡嗡叫
duplicate ['duːplikeit] vt. 复制；复印；复写
durable ['djuərəbl] adj. 持久的；耐用的
dumpster ['dʌmpstə] n. 大垃圾桶
dwindle ['dwindl] v. 缩小
dynamical [dai'næmikəl] adj. 动力（学）；有力量的
dynamics [dai'næmiks] n. 动力学
easting ['iːstiŋ] n. 东西距；朝东方；东行航程
electromagnetic [iˌlektrəumæg'netik] adj. 电磁的
　　electronic conductor 电子导电体
eliminate [i'limineit] vt. 消除；排除
ellipsoid [i'lipsɔid] n. 椭圆体
elongate ['iːlɔŋgeit] v. 拉长；（使）伸长；延长
embrace [im'breis] v. 拥抱；互相拥抱；包含；收买；信奉 n. 拥抱
emit [i'mit] v. 发出；散发（辐射、光、热、气等）；发行
empirical [em'pirikəl] adj. 完全根据经验的；经验主义的
empower [im'pauə] v. 授权；给……的权力；使能够
encoder [in'kəudə] n. 编码器；译码器
encounter [in'kauntə] v. 遇到；面临
encyclopedia [inˌsaiklə'piːdiə] n. 百科全书
endogenic [ˌendəu'dʒenik] adj. ［地］内成的；内生的；内营力的
endurance [in'djuərəns] n. 忍耐力；耐久性
enforce [in'fɔːs] vt. 强迫；执行；坚持；加强
entertain [ˌentə'tein] v. 给……娱乐；使快乐；对（意见、建议、感情）给予考虑
entity ['entiti] n. 实体
entrepreneurship [ˌɔntrəprə'nəːʃip] n. 企业家［主办人等］的身份［地位、职权、能力］
ephemeris [i'femaris] n. ［天］星历表；历书
epoch ['iːpɔk, 'epɔk] n. 新纪元；时代；时期；时间上的一点；［地质］世
equally-spaced ['ikwəli speist] adj. 等距的；间距相等的；同样间隔的
equator [i'kweitə] n. 赤道；赤道线
equidistant [ˌiːkwi'distənt] adj. 距离相等的；等距的

205

equilibrium [ˌi:kwi'libriəm] n. 平衡；平静；均衡；保持平衡的能力；沉着；安静
equipotential [ˌi:kwipə'tenʃəl] adj. 等位的；有相等潜力的；等电位的
equivalence [i'kwivələns] n. 同等；[化]等价；等值
equivalent [i'kwivələnt] adj. 相等的；相同的；等同的 n. 对等的人；当量
essay ['esei] n. 论说文；散文；随笔；小品文
Euclidean Space 欧几里得空间
evaluate [i'væljueit] vt. 评价；估计；求……的值
evenly ['i:vənli] adv. 均匀地；平坦地
evidence ['evidəns] n. 根据；证明；迹象；痕迹；证据 v. 证实；证明
evolve [i'vɔlv] v. (使)发展；(使)进展；(使)进化
excavation [ˌekskə'veiʃən] n. 挖掘；发掘；挖掘成的洞；出土文物
exclude [ik'sklu:d] vt. 拒绝接纳；把……排除在外；排斥
exert [ig'zə:t] vt. 施加(压力)；努力；发挥；竭尽全力
exhaustive [ig'zɔ:stiv] adj. 无遗漏的；彻底的；详尽的；无遗的
exogenic [ˌeksəu'dʒenik] adj. [生]外生的；[地]外成的；[医]外因的
expenditure [ik'spenditʃə] n. 花费；支出；开支
expertise [ˌekspə'ti:z] n. 专门技术；专家的意见
explicit [ik'splisit] adj. 清楚的；外在的；坦率的；(租金等)直接付款的
exposition [ˌekspə'ziʃn] n. 阐述；展览会；说明文
  expository essay 说明文
exposure [ik'spəuʒə] n. 曝光；暴露；揭露；面临
exterior [ik'stiəriə] adj. 外部的；外面的；外表的；户外的 n. 外观；外表
falsify ['fɔ:lsifai] v. 伪造
falsification [ˌfɔ:lsifi'keiʃən] n. 歪曲；弄虚作假；串改；伪造
federation [ˌfedə'reiʃən] n. 联合；同盟；联邦；联盟
fidelity [fi'deliti] n. 忠实；诚实；忠诚；保真度；逼真度；保真度；重现精度
fieldwork ['fi:ldwə:k] n. 野外工作；实地调查
filter ['filtə] n. 滤波器；过滤器；滤光器；筛选
flattening ['flætniŋ] n. 扁率；整平
fledgling ['fledʒliŋ] n. 初出茅庐的人；无经验的人；刚会飞的幼鸟
flexibility [ˌfleksə'biliti] n. 适应性；机动性；挠性
flock [flɔk] n. 群；一大群人 v. 聚集；群集；蜂拥
font [fɔnt] n. 字型；一幅活字；洗礼盆；盛油罐
foresight ['fɔ:sait] n. 前视；远见；深谋远虑
forestry ['fɔristri] n. 林产；森林地；林学
forum ['fɔ:rəm] n. 论坛；法庭；讨论会
frame [freim] vt. 构成；设计；制定
framework ['freimwə:k] n. 构架；框架；结构
  free station 自由设站
fundamental [ˌfʌndə'mentl] n. 基本原则；基本原理

Gal *n.* [加速度单位]伽；[重力加速度单位]加仑

Galileo [ˌgæli'leiəu] *n.* 伽利略(1564—1562，意大利物理及天文学家)

gear [giə] *n.* 齿轮；传动装置

geocoding [ˌdʒiəu'kəudiŋ] *n.* 地理编码

geodynamic [ˌdʒi(:)əudai'næmik] *adj.* 地球动力学的

geographical [ˌdʒiə'græfikəl] *adj.* 地理学的；地理的

geoid ['dʒi:ɔid] *n.* [地]大地水准面

geoidal ['dʒi:ɔidl] *adj.* 大地水准面的

geology [dʒi'ɔlədʒi] *n.* 地质学；地质概况

　　geological fault 地质断层

geomagnetic [ˌdʒi:əumæg'netik] *adj.* 地磁的；地磁学的

geomagnetism [ˌdʒ(i:)əu'mægnitizəm] *n.* 地磁；地磁学

geometric [dʒiə'metrik] *adj.* 几何的；几何学的

geophysics [ˌdʒi(:)əu'fiziks] *n.* 地球物理学

geoscience [ˌdʒi(:)əu'saiəns] *n.* 地球科学

geospatial [ˌdʒiəu'speiʃəl] *adj.* 地理空间的

geostationary [ˌdʒiəu'steiʃənəri] *adj.* 与地球旋转同步的；与地球同步的

geotechnical [dʒi:əu'teknikəl] *adj.* 岩土工程技术的

gerund ['dʒerənd] *n.* 动名词

grab [græb] *v.* 攫取；抓取；吸引……的注意力；使留下印象

graduated ['grædjueitid] *adj.* 分度的；分级的

graticule ['grætikju:l] *n.* 分成小方格以便复制的图形；格子线

gravitation [ˌgrævi'teiʃən] *n.* 地心吸力；引力作用

gravity ['græviti] *n.* 重力；地心引力

　　gravity field 重力场

grinder ['graində] *n.* 磨工；尤指磨刀具的工人；研磨者

gutter ['gʌtə] *n.* 水槽；檐槽；排水沟；槽；贫民区

handheld ['hændhɛld] *adj.* 手持型；掌上型

Harvard ['hɑ:vəd] *n.* 美国哈佛大学

Hawaii [hɑ:'vaii:] *n.* 夏威夷；夏威夷岛

hazard ['hæzəd] *n.* 冒险；危险；冒险的事

hazardous ['hæzədəs] *adj.* 危险的；冒险的；碰运气的

headquarters ['hedˌkwɔ:təz] *n.* 司令部；指挥部；总部

herald ['hɛrəld] *v.* 预示；是……的前兆；宣布 *n.* 预兆；使者

hinder ['hində] *v.* 阻碍；打扰

homogeneous [ˌhɔmə'dʒi:njəs] *adj.* 同类的；相似的；均一的；均匀的

horizon [hə'raizn] *n.* 地平线；地平(线)；范围；视野

hybrid ['haibrid] *n.* 混合物 *adj.* 混合的；杂交成的

hydroelectric ['haidrəui'lektrik] *adj.* 水力电气的

hydrographic [ˌhaidrəu'græfik] *adj.* 与水道测量有关的；与水文地理有关的

hydrographic survey 海道测量；水道测量
hyperspectral [ˌhaipəˈspektrəl] adj. 高光谱的
identify [aiˈdentifai] vt. 确定；识别，鉴别
IEC 国际电工委员会
illustrative [ˈiləstreitiv] adj. 说明性的；例证性的
imagery [ˈimidʒəri] n. 肖像(总称)；雕刻影像
immerse [iˈməːs] v. 浸；投入；陷入；沉浸于
implicitly [imˈplisitli] adv. 含蓄地；暗中地
inconsistency [ˌinkənˈsistənsi] n. 矛盾
incorporation [inˌkɔːpəˈreiʃən] n. 结合；合并；形成法人组织；组成公司(或社团)
indebtedness [inˈdetidnis] n. 受恩惠；亏欠；债务
indexing [ˈindeksiŋ] n. 标定指数
indication [ˌindiˈkeiʃən] n. 指出；指示；迹象；暗示
indicator [ˈindikeitə] n. 指示器；[化]指示剂
indices [ˈindisiːz] n. index 的复数，[数学]指数；指标；(刻度盘上)指针
industrialization [inˌdʌstriəlaiˈzeiʃ(ə)n] n. 产业化；工业化
infer [inˈfəː] v. 推断
inflecxion [inˈflekʃən] n. [数] 拐点
informatics [ˌinfəˈmætiks] n. 信息学；情报学
infrared [ˌinfrəˈred] adj. 红外线的 n. 红外线
infrastructure [ˈinfrəstrʌktʃə] n. 基础下部组织；下部构造
initialize [iˈniʃəlaiz] vt. 初始化
innovation [ˌinəuˈveiʃən] n. 改革；创新
innovative [ˈinəuˈveitiv] adj. 创新的；革新(主义)的
innovator [ˈinəuveitə] n. 改革者；革新者
instantaneous [ˌinstənˈteinjəs] adj. 瞬间的；即刻的；即时的
instill [inˈstil] vt. 逐渐灌输；徐徐滴入
instinctively [inˈstiŋktivli] adv. 本能地
instrumentation [ˌinstrumentˈteiʃən] n. 使用仪器
integral [ˈintigrəl] adj. 必需的；必要的；基本的；基础的
integrity [inˈtegriti] n. 正直；诚实；完整；完全；完整性
intent [inˈtent] n. 意图；目的；意向
intercept [ˌintəˈsept] vt. 截取；中途阻止
interdisciplinary [ˌintə(ː)ˈdisiplinəri] adj. 各学科间的
interface [ˈintəfeis] n. 接合点；结合点；边缘区域；界面；接口
interference [ˌintəˈfiərəns] n. 干扰；冲突；干涉
interferometer [ˌintəfiəˈrɔmitə] n. 干涉计
interferometric [ˈintəˌferəuˈmetrik] adj. 干涉仪的；用干涉仪测量的
interferometry [ˌintəfiəˈrɔmitri] n. 干涉测量(法)
intermediate [ˌintəˈmiːdjət] adj. 中间的

interior [in'tiəriə] *adj.* 内部的；里面的；国内的；内政的 *n.* 内部；里面；内陆
internal [in'tə:nl] *adj.* 内在的；国内的
interpolation [in,tə:pə'leiʃən] *n.* 插补；内插法；插值
interpretation [in,tə:prəteiʃən] *n.* 解释；阐明；口译；通译
intersect [,intə'sekt] *vi.* （直线）相交；交叉
intervene [,intə'vi:n] *vi.* 干涉；干预；插入；介入；（指时间）介于其间 *v.* 干涉
intervention [,intə'vɛnʃən] *n.* 干预；介入；调解
interview ['intəvju:] *n.* 接见；会见；采访；面试 *v.* 采访；询问；进行面试；审问；表现
introductory [,intrə'dʌktəri] *adj.* 介绍性的
intuitively [in'tju(:)itivli] *adv.* 直觉地；直观地
invar [in'vɑ:] *n.* 因瓦；不胀钢
invariably [in'vɛəriəb(ə)li] *adv.* 不变地；总是
investigation [in,vesti'geiʃən] *n.* 调查；研究
ionosphere [ai'ɔnəsfiə] *n.* 电离层
irregular [i'regjulə] *adj.* 不规则的；无规律的
ISO(International Standardization Organization) *abbr.* 国际标准化组织
isometric [,aisə'metrik] *adj.* 等大的；等容积的
ISSN *abbr.* 国际标准期刊编号
iterative ['itərətiv] *adj.* [数]迭代的；重复的；反复的
keen [ki:n] *adj.* 锋利的；敏锐的；敏捷的；热心的；渴望的
kinematic [,kinə'mætik] *adj.* 非静止的；运动学的；运动学上的
latitude ['lætitju:d] *n.* 纬度；范围；（复数）地区
lattice ['lætis] *n.* 花格结构；格构；交错结构
    law of cosine 余弦定律
    law of sines 正弦定律
layout ['lei,aut] *n.* 放样；规划；设计；（工厂等的）布局图
legislation [,ledʒis'leiʃən] *n.* 立法；法律的制定（或通过）
lift [lift] *n.* （空气的）升力；电梯；举；抬；提 *v.* 举起；提高；解除；消散
lining ['lainiŋ] *n.* 衬里；内胆；里子
literally ['litərəli] *adv.* 照字面意义；逐字地
logistics [lə'dʒistiks] (*pl.*)-*n.* 后勤；物流；货物配送
longitude ['lɔndʒitju:d] *n.* 经度；经线经度
loxodrome ['lɔksədrəum] *n.* 斜航线；恒向线
magma ['mægmə] *n.* 岩浆；[药]乳浆剂
magnetosphere [mæg'ni:təsfiə] *n.* 磁气圈
magnitude ['mægnitju:d] *n.* 数量；巨大；广大；量级
maintenance ['meintinəns] *n.* 维护；保持；生活费用；扶养
maneuverability [mə,nuvərə'biliti] *n.* [航][车辆][船]机动性；可操作性
manipulate [mə'nipjuleit] *vt.* （熟练地）操作；使用（机器）；操纵（人或市价）；利用
manufacturer [,mænju'fæktʃərə] *n.* 制造业者；厂商

  map projection 地图投影
margin ['mɑːdʒin] n. 页面的空白边；页边线；边；边沿；边缘；数量；利润；定金；押金
marksman ['mɑːksmən] n. 射手；神射手
Massachusetts [ˌmæsə'tʃuːsits] n. 马萨诸塞州
mean [miːn, min] n. 平均数；中间；中庸
mechanical [mi'kænikl] adj. 机械的；机械制的；机械似的；呆板的
membership ['membəʃip] n. 会员资格；会员身份
Mercator [məːˈkeitɔː] n. 墨卡托(1512—1594，荷兰佛兰德斯的地理学家、地图制作家)
merge [mɜːdʒ] v. (使)融合；(使)合并；融入
meridian [məˈridiən] n. 子午线；正午；顶点；全盛时期 adj. 子午线的；正午的
  prime meridian 本初子午线；本初子午圈线
mesh [meʃ] n. 格网；网孔；网丝；网眼；圈套；陷阱
methodology [ˌmeθəˈdɔlədʒi] n. 方法学；方法论
meticulous [məˈtikjələs] adj. 小心翼翼的
micrometer [maiˈkrɔmitə] n. 测微计；千分尺
microprocessor [ˌmaikrəuˈprəusesə] n. [计]微处理器
microsatellite [ˈmaikrə(u)ˌsætəlait] n. 微卫星
milestone [ˈmailstəun] n. 里程碑；里程标；重要事件；转折点
mill [mil] n. 工厂；制造厂
mindset [ˈmaindset] n. 心理定势；心态
miniaturization [ˌminətʃəraiˈzeiʃn] n. 微型化；小型化
miniaturize [ˈminiətʃəraiz] vt. 使小型化
miscellaneous [ˌmisəˈleinjəs] adj. 各色各样混在一起；混杂的；多才多艺的
mislead [misˈliːd] vt. 误导
mitigation [ˌmitiˈgeiʃən] n. 缓解；减轻；平静
moderately [ˈmɔdəreitli] adv. 适度地
modulated [ˈmɔdjuleitid] adj. 已调整[制]的；被调的
module [ˈmɔdjuːl] n. 模数；模块；登月舱；指令舱
moisture [ˈmɔistʃə] n. 湿度；湿气；潮湿
mole [məul] n. 全断面掘进机
  tunneling mole 隧道掘进机
monitor [ˈmɔnitə] vt. 监控；n. 监测；监视；控制；追踪；监控器
monograph [ˈmɔnəugrɑːf] n. 专论
monument [ˈmɔnjumənt] n. 纪念碑
  permanent monument 永久标石
  monumentation 埋石
monomer [ˈmɔnəmə] n. 单体；单元结构
morphology [mɔːˈfɔlədʒi] n. [生物]形态学；形态论；[语法]词法、词态学
  mathematical morphology 数学形态学
motivation [ˌməutiˈveiʃn] n. 动机；动力；诱因

motorize ['məutəraiz] vt. 使机动化；使摩托化

mould [məuld] v. 铸造；用土覆盖

MTBF abbr. ［军］Mean Time Between Failures，平均故障间隔时间；［计］平均故障间隔时间

multiply ['mʌltiplai] v. 乘；增加；繁殖

multirotor [ˌmʌlti'rəutə] adj. 有多个转子的；有多旋翼的　n. 多轴飞行器

multispectral [ˌmʌlti'spektrəl] adj. 多光谱的；多谱的

multistep ['mʌltiˌstep] adj. 多步的；多级的

Munich ['mju:nik] n. 慕尼黑(德国城市，巴伐利亚州首府)

mutually ['mju:tʃuəli] adv. 互相地；互助地

National Academy of Science 国家科学院

narration [nə'reiʃn] n. 叙述；故事；讲述；记叙文
    narrative essay 记叙文

navigation [ˌnævi'geiʃən] n. 导航；航海；航空；领航；航行

necessitate [ni'sesiteit] v. 成为必要

nevertheless [ˌnevəð'les] conj. 然而；不过

Newton ['nju:tn] n. 牛顿

nickel ['nikl] n. ［化］镍；镍币；(美国和加拿大的)五分镍币

node [nəud] n. 节点

nominal ['nɔminl] adj. 名义上的；有名无实的；名字的；［语］名词性的

noncontact [ˌnɔn'kɔntækt] n. 无接点　adj. 非接触(式)的；无接触的

non-fictional [nɔn'fikʃənl] adj. 非虚构叙的

nonparametric ['nɔnpærəm'etrik] adj. 非参数的

non-singular ['nəun-siŋgjələ] adj. 非奇异的

norm [nɔ:m] n. 规范；准则

normal ['nɔ:məl] n. 正态；正规；常态；［数］法线

northing ['nɔ:θiŋ] n. 北距(向北航行的距离)；北进；北航

notable ['nəutəbl] adj. 值得注意的；显著的

notarize ['nəutəraiz] v. 由公证人确认……合法；公证
    notarized certificate 公证证明

noun [naun] n. 名词

oblate ['ɔbleit] adj. ［数］扁平的；扁圆的

oblateness ['ɔbleitnis] n. 扁率；扁球形；扁圆形

oblique [ə'bli:k] adj. 倾斜的；间接的

obstacle ['ɔbstəkəl] n. 障碍；阻碍；障碍物

obviate ['ɔbvieit] vt. 消除；排除(危险、障碍)；回避；预防；避免

occlusion [ə'klu:ʒən] n. 闭塞；阻塞；遮挡

oceanography [ˌəuʃiə'nɔgrəfi] n. 海洋学

odometer [əu'dɔmitə] n. ［美］(汽车等的)里程表；自动计程仪

offset ['ɔ:fset] n. 偏移量；抵销；弥补；分支

on-site 场区内；厂区内；在工地上
ophthalmic [ɔf'θælmik] *adj.* 眼的；眼科的
optimal ['ɔptiməl] *adj.* 最佳的；最理想的
optimization [ˌɔptimai'zeiʃən] *n.* 最佳化；最优化
optimum ['ɔptiməm] *adj.* 最适宜的
orientation [ˌɔ(ː)riən'teiʃən] *n.* 定位；方向；方位；倾向性；向东方
origin ['ɔridʒin] 原点；起算点
orthogonal [ɔː'θɔgənəl] *adj.* 正交的；直角的；直交的
orthographic [ˌɔːθə'græfik] *adj.* 正射的；正射投影的；[语]正字法的；拼字正确的
orthometric [ˌɔːθə'metrik] *adj.* 正高的
orthophoto [ˌɔːθəu'fəutəu] *n.* [摄]正射影像；正射像片
ortho-rectification [ˌɔːθəu-ˌrektifi'keiʃn] *n.* 正射纠正；正射校正
outreach [aut'riːtʃ] *v.* 到达顶端；超越
overambitious ['əuvəræm'biʃəs] *adj.* 野心太大的
overlay [ˌəuvə'lei] *n.* 覆盖；覆盖图
overlook [ˌəuvə'luk] *vt.* 没注意到；俯瞰；耸出；远眺
Pacific Ocean 太平洋
painstakingly ['peinsteikiŋli] *adv.* 辛苦地；辛勤地；艰苦地
paleomagnetism [ˌpæliəu'mægnətizəm] *n.* [物]古磁学
panchromatic [ˌpænkrəu'mætik] *adj.* [摄]全色的
parameter [pə'ræmitə] *n.* 参数；参量
parcel ['pɑːsl] *n.* 小包；包裹
parent company 母公司；总公司
particle ['pɑːtikl] *n.* 粒子；点；极小量；微粒；质点
paving ['peiviŋ] *n.* 衬砌，铺路
payload ['peiləud] *n.* (飞机、船只的) 载荷；有效负荷；装载量
payment ['peimənt] *n.* 付款；支付；报酬；偿还；报应；惩罚
pedestrian [pə'destriən] *n.* 步行者；行人
pendulum ['pendjələm] *n.* 钟摆；摇锤
penetration [ˌpenə'treiʃ(ə)n] *n.* 穿透；渗透；进入
perception [pə'sɛpʃən] *n.* 感觉；感知；看法；洞察力
periodicity [ˌpiəriə'disiti] *n.* [数]周期性；频率；定期出现
permanent ['pəːmənənt] *adj.* 永久的；持久的
perpendicular [ˌpəːpən'dikjulə] *adj.* 垂直的；正交的
perspective [pə'spektiv] *n.* 透视画法；透视图；远景；观察
persuade [pə'sweid] *v.* 说服；劝服；敦促；使相信；使信服；驱使
pertaining [pə(ː)'teiniŋ] *adj.* 与……有关系的；附属……的；为……固有的(to)
phenomena [fi'nɔminə] *n.* 现象
philosophy [fi'lɔsəfi] *n.* 哲学；哲学体系；达观；冷静
photogrammetry [ˌfəutəu'græmitri] *n.* [测]摄影测量学；摄影测量法

photography [fə'tɔgrəfi] n. 摄影；摄影术；照相术
pitch [pitʃ] n. [航]俯仰；沥青；球场；场地；颠簸；倾斜度
pixel ['piksəl] n. 像素
planar ['pleinə] adj. 平面的；平坦的
planetary ['plænitri] n. 行星的
plotter ['plɔtə] n. 绘图仪
plotting ['plɔtiŋ] n. 标图；测绘
plumb [plʌm] n. 铅锤；铅弹 adj. 垂直的 vt. 使垂直；探测
    plumb line 铅垂线
polygon ['pɔligən] n. [数]多角形；多边形
polynomial [ˌpɔli'nəumiəl] adj. [数]多项式的
populate ['pɔpjuleit] v. 使居住；增殖；粒子数增加
portable ['pɔːtəbl] adj. 轻便的；手提(式)的；便携式的
portal ['pɔːtl] n. 门；入口
portray [pɔː'trei] v. 描绘
potential [pə'tenʃ(ə)l] n. 潜能；潜力；电压
practitioner [præk'tiʃənə] n. 从业者；开业者
preclude [pri'kluːd] n. 排除
predetermine [ˌpriːdi'təːmin] v. 预定；预先确定
predictable [pri'diktəbəl] adj. 可预测的；可预见的；意料之中的
predominantly [pri'dɔminəntli] adv. 绝大多数；主要地；占主导地位地
pre-dredge [pri-'drɛdʒ] v. 疏浚；挖掘
prefabricate [pri'fæbrəˌkeit] v. 预先建造；预制
preliminary [pri'liminəri] adj. 初步的；预备的
prerequisite ['priː'rekwizit] n. 先决条件
preservation [ˌprezə(ː)'veiʃən] n. 保存
prestigious [pre'stidʒəs] adj. 享有声望的；声望很高的
privilege ['privəlidʒ] n. 特权；特惠待遇；优待；特别待遇
prism ['prizəm] n. [物]棱镜
probability [ˌprɔbə'biliti] n. 概率；可能性；或然性
    probability curve 概率曲线；或然率曲线
product ['prɔdʌkt] n. 乘积；产品；产物
prohibitive [prə'hibitiv] adj. 禁止的；抑制的
projection [prə'dʒekʃən] n. 投影；投射；投影图；地图投影；抛；规划
prompt [prɔmpt] v. 引起；激起；鼓励……往下说；追问；提示；促使……采取行动
propagate ['prɔpəgeit] v. 传播；宣传
    error propagation 误差传播
propel [prə'pɛl] v. 推进；推动；驱使；驱动
property ['prɔpəti] n. 所有物；所有权；性质；特性；(小)道具
proposal [prə'pəuzl] n. 提议；建议；建议书

proposition [ˌprɔpəˈziʃən] n. 命题；主张；建议；陈述
proprietor [prəˈpraiətə] n. 所有者；经营者
prospective [prəˈspektiv] adj. 预期的；盼望中的；未来的；未来可能发生的
protractor [prəˈtræktə] n. 量角器
provision [prəˈviʒən] n. 供应；(一批)供应品；预备；防备；规定
proximal [ˈprɔksiməl] adj. 最接近的
pseudo [ˈsjuːdəu] adj. 假的；冒充的
pseudorange [ˈsjuːdəureidʒ] n. 伪距
quadrant [ˈkwɔdrənt] n. 象限；四分仪
qualification [ˌkwɔlifiˈkeiʃən] n. 资格；条件；限制；限定；赋予资格
quantify [ˈkwɔntifai] vt. 量化；确定数量
quarterly [ˈkwɔːtəli] adj. 一年四次的；每季的
radiation [ˌreidiˈeiʃən] n. 辐射的热(或能量等)；放射线；辐射形进化；放射疗法
radius [ˈreidiəs] n. 半径；范围；辐射光线；有效航程；界限
rarely [ˈrɛəli] adv. 很少地；罕有地
raster [ˈræstə] n. [物]光栅；栅格
rationale [ˌræʃəˈnɑːl] n. 基本原理
real estate 房地产；房地产所有权
receipt [riˈsiːt] n. 收到；收条；收据
recipient [riˈsipiənt] n. 接受者；收受者
reciprocal [riˈsiprəkəl] adj. 彼此相反的；互惠的；相应的；倒数的
recommendation [ˌrekəmenˈdeiʃən] n. 推荐；介绍(信)；劝告；建议
reconnaissance [riˈkɔnisəns] n. 勘测；侦察；搜索
recruiter [riˈkruːtə] n. 招聘人员；征兵人员
rectangular [rekˈtæŋgjulə] adj. 矩形的；成直角的
redundancy [riˈdʌndənsi] n. 冗余
reestablish [ˌriːiˈstæbliʃ] v. 重建；使复原；使复位
referee [ˌrefəˈriː] n. 仲裁人；调解人；[体]裁判员
reference [ˈrefrəns] n. 引证；证明信；介绍信；引文
refined [riˈfaind] adj. 精制的；优雅的；精确的
refraction [riˈfrækʃən] n. 折光；折射
regards [riˈgɑːdz] n. (用于信函结尾或转达问候)致意；问候
registration [ˌredʒisˈtreiʃən] n. 注册；报到；登记
reinforcement [ˌriːinˈfɔːsmənt] n. 加固物；增强材料
relief [riˈliːf] n. 地貌；地势；地势起伏；地形
render [ˈrendə] vt. 致使；呈递；归还；汇报；放弃；表演；实施
repository [riˈpɔzitəri] n. 贮藏室；智囊团；知识库；仓库
representative [ˌrepriˈzentətiv] n. 代表 adj. 典型的；有代表性的
reproducibility [ˌriprəˌdjuːsiˈbiliti] n. 重复能力；再现性
residual [riˈzidjuəl] adj. 剩余的；残留的

resultant [ri'zʌltənt] *adj.* 作为结果而发生的；合成的
résumé (resume) ['rezju:mei; ri'zju:m; 'rezəmei] *n.* 简历
retrieval [ri'tri:vəl] *n.* 检索；恢复；修补；重获
revenue ['revinju:] *n.* 收入；国家的收入；税收
reverse [ri'vɜ:s] *adj.* 反向的；倒转的；颠倒顺序的
  reverse-chronological 倒序的；按时间逆序排列的
revolution [ˌrevə'lu:ʃən] *n.* 旋转；革命
rhumb [rʌm] *n.* 罗盘方位；罗盘方位单位
  rhumb line 等角线；恒向线；无变形线；斜航
ridge [ridʒ] *n.* 山脊；屋脊；山脉；犁垄
  ridge line 屋脊线
rigorously ['rigərəsli] *adv.* 严密地；严厉地；残酷地；严格地
roll [rəul] *n.* [航]翻滚，横滚；卷；滚动 *v.* (使)翻滚；滚动
rotation [rəu'teiʃən] *n.* 旋转
rover ['rəuvə] *n.* 流浪者；漫游者；自由队员
salutation [ˌsælju'teiʃn] *n.* 致意；(信函中如 Dear Sir 之类的)称呼语
satisfactory [ˌsætis'fæktəri] *adj.* 满意的；赎罪的
scatter ['skætə] *v.* 分散；散开；撒开；驱散
secant ['si:kɔnt] *adj.* 切的；割的；交叉的 *n.* 割线；正切
seep [si:p] *v.* 渗出；渗漏
seismology [saiz'mɔlədʒi] *n.* 地震学
selenodesy [ˌseli'nɔdisi] *n.* [天]月面测量学
semimonthly [ˌsemi'mʌnθli] *adj.* 每半个月一次的 *n.* 半月刊
sensitivity [ˌsensi'tiviti] *n.* 灵敏(度)；灵敏性
sensor ['sensə] *n.* 传感器
servo ['sə:vəu] *n.* 伺服；伺服系统
sewer ['sjuə] *n.* 下水道；缝具；缝纫者
sexagesimal [ˌseksə'dʒesiməl] *adj.* 六十的；六十进位的
  sexagesimal system 六十分制
sextant ['sekstənt] *n.* 六分仪
shotcrete ['ʃɔtkri:t] *n.* 压力喷浆；喷射的水泥砂浆
shoulder ['ʃəuldə] *vt.* 肩负；承当
showcase ['ʃəukeis] *v.* 展示；表现 *n.* 玻璃陈列柜；展示场所
signature ['signətʃə] *n.* 签名；署名
significantly [sig'nifikəntli] *adv.* 意味深长地；值得注目地
simplification [ˌsimplifi'keiʃən] *n.* 简化
simulation [ˌsimju'leiʃən] *n.* 仿真；假装；模拟
simultaneously [ˌsiməl'teiniəsli] *adv.* 同时地
sincerely [sin'siəli] *adv.* 真诚地；诚实地
singular ['siŋgjulə] *adj.* 奇异的；单一的；非凡的；持异议的

slide [slaid] *n.* 滑；滑动；幻灯片
solar-terrestrial relation 日地关系
sophisticate [sə'fistikeit] *vt.* 弄复杂；篡改；使变得世故；掺合
sophistication [sə,fisti'keiʃən] *n.* 复杂；强词夺理；诡辩；混合
spaceborne ['speisbɔ:n] *adj.* 宇宙飞行器上的；卫星[飞船]上的
spacecraft ['speiskrɑ:ft] *n.* 太空船
sparse [spɑ:s] *adj.* 稀少的；稀疏的
sparsity ['spɑ:səti] *n.* 稀少
specialization [,speʃ(ə)li'zeiʃ(ə)n] *n.* 专业；专门化；特殊化
  interdisciplinary specialization 交叉学科专业
specification [,spesifi'keiʃən] *n.* 详述；规格；说明书；规范
spectral ['spεktrəl] *adj.* [光]光谱的；幽灵似的；鬼怪的
spherical ['sferikəl] *adj.* 球形的；球的
spheroid ['sfiərɔid] *n.* 球状体；回转椭圆体
spin [spin] *v.* 旋转；纺；纺纱
spindle ['spindl] *n.* 轴；杆；心轴；锭子；纺锤
stadia ['steidjə] *n.* 视距；视距仪器
standpoint ['stændpɔint] *n.* 立场；观点
  state-of-the-art 发展现状；技术发展现状
statistical [stə'tistikəl] *adj.* 统计的；统计学的
  mathematical statistics 数理统计
static ['stætik] *adj.* 静的；静态的；静止的；不动的
statute ['stætʃu:t] *n.* 法令；条例
stereoplotting [,steriə'plɔtiŋ] *n.* 立体测图
stitch [stitʃ] *v.* 缝；缝补；缝合 *n.* 针脚；缝线；缝纫法
stochastic [stə'kæstik] *adj.* 随机的
  stochastic variable, random variable 随机变量
stock [stɔk] *n.* 股票；股份；托盘；祖先；血统；原料
straightforward [,streit'fɔ:wəd] *adj.* 直截了当的；简单的；易懂的；正直的；坦率的
subdiscipline ['sʌb'disiplin] *n.* (学科的)分支；分科
subjective [səb'dʒektiv] *n.* 对象；主题；问题；主语 *adj.* 主观的
subscribe [səb'skraib] *v.* 订阅；订购；认购；支付
subsidence ['sʌbsidəns] *n.* 下沉；沉淀；陷没
subsidiary [səb'sidjəri] *adj.* 辅助的；补充的
substantially [səb'stænʃ(ə)li] *adv.* 充分地
subterranean [,sʌbtə'reiniən] *adj.* 地面下的；在地下进行的
suburban [sə'bə:bən] *adj.* 郊外的；偏远的
succinct [sək'siŋkt] *adj.* 简洁的；简明的；简练的
sunken ['sʌŋkən] *adj.* 沉没的；没入水中的；凹陷的；下陷的；深陷的
sun-seeking ['sʌnsi:kiŋ] *a.* 太阳定向

superfluous [ˌsuːˈpɜːfluəs] *adj.* 多余的；过剩的；过量的
supersede [ˌsjuːpəˈsiːd] *vt.* 代替；取代；接替；紧接着……而来；[律]延期
supervisor [ˈsuːpəvaizə] *n.* 研究生导师；监督者
suspect [sʌsˈpekt] *v.* 怀疑；猜想；对……有所觉察
sustain [səˈstein] *v.* 支持；支撑；忍受；经受
Swiss [swis] *adj.* 瑞士的；瑞士人的
swoop [swuːp] *n.* 突然下降；猛扑
symbology [simˈbɔlədʒi] *n.* 符号学；象征手法
symposia [simˈpəuziə] *n.* 座谈会；评论集
synthetic [sinˈθetik] *adj.* 合成的；人造的；综合(型)的 *n.* 合成物；合成纤维(织物)
tacheometry [ˌtækiˈɔmitri] *n.* [测]视距测量(法)
tackle [ˈtækl] *v.* 决心处理；与……交涉；与……磋商
tailor [ˈteilə] *vt.* 专门制作；订做
  tailor to the situatio 根据情况量身定做
tandem [ˈtændəm] *n.* 串联；双人自行车；并行 *adv.* 同时地
tangent [ˈtændʒənt] *adj.* 相切的；切线的 *n.* 切线；[数]正切
tectonic [tekˈtɔnik] *adj.* [建]构造的；建筑的
tectonics [tekˈtɔniks] *n.* 筑造学
tectonophysics [ˌtektənəuˈfiziks] [复数] *n.* [用作单数] [物]地壳构造物理学
telecommunication [ˈtelikəmjuːniˈkeiʃən] *n.* 电信；长途通信；无线电通信；电信学
telescope [ˈteliskəup] *n.* 望远镜
temporal [ˈtempərəl] *adj.* 时间的；当时的；暂时的；现世的；世俗的
tenure [ˈtenjə] *n.* 保有；保有条件；占有；任职；终身职位；任期
  land tenure 土地使用权；土地所有制；土地占有制
terminate [ˈtəːmineit] *v.* 停止；结束；终止
terminology [ˌtəːmiˈnɔlədʒi] *n.* 术语学
terrain [ˈtərein] *n.* 地形
terrestrial [təˈrestriəl] *adj.* 陆地的
textile [ˈtekstail] *n.* 纺织品 *adj.* 纺织的
texture [ˈtekstʃə] *n.* 纹理；质地；口感
theodolite [θiˈɔdəlait] *n.* [测]经纬仪
theoretical [θiəˈretikəl] *adj.* 理论的
therewith [ðɛəˈwiθ] *adv.* 与此；以其；于是
thermal [ˈθəːməl] *adj.* 热的；热量的；(衣服)保暖的 *n.* 上升的热气流
thumb [θʌm] *n.* 拇指
  rules of thumb 经验法则
tilt [tilt] *v.* 倾斜；侧倾 *n.* 倾斜；侧倾；倾向；偏向；倾斜面；斜坡；俯仰转动
tolerance [ˈtɔlərəns] *n.* 公差；容限；限差
topographic [ˌtɔpəˈgræfik] *adj.* 地志的；地形学上的
topological [təˈpɔlədʒikl] *adj.* 拓扑的

tough [tʌf] *adj.* 艰苦的；艰难的；棘手的
trademark ['treidmɑ:k] *n.* 商标
trajectory ['trædʒiktəri] *n.* [物]（射线的）轨道；弹道；轨线
transcend [træn'send] *v.* 超出；超越；胜过；超过
transcript ['trænskript] *n.* 学生成绩报告单；文本；转录物
translation [trænz'leiʃən] *n.* [数]平移；翻译；译文；转换[物]
transverse ['trænzvə:s] *adj.* 横向的；横断的
  Universal Transverse Mercator (UTM) 通用横墨卡托投影
traverse ['trævə(:)s] *n.* 导线；横贯；横断
  traversing 导线测量
treatise ['tri:tiz] *n.* 论文；论述
tremendous [tri'mendəs] *adj.* 极大的；巨大的
trench [trentʃ] *n.* 沟渠；堑壕；管沟；电缆沟；战壕
  trial and error 累试法
triangle ['traiæŋgl] *n.* [数]三角形；三人一组；三角关系
triangulateration [trai͵æŋgju'leitəreiʃən] *n.* 边角测量
triangulation [͵traiæŋgju'leiʃən] *n.* 三角测量
trigger ['trigə] *vt.* 引发；引起；触发
trigonometry [͵trigə'nɔmitri] *n.* 三角法
  plane trigonometry 平面三角
trilateration [trai͵lætə'reiʃən] *n.* 三边测量
trispectral [trai'spektrəl] *n.* 三色谱
tropic ['trɔpik] *n.* [天]（天球的）回归线；热带
turnaround ['tɜ:nə͵raund] *n.* 周转；转变；转向；好转
tutorial [tju:'tɔ:riəl] *n.* 指南 *adj.* 导师的
U. S. Department of Defense (DoD) 美国国防部
unbiased [ʌn'baiəst] *adj.* 无偏的；没有偏见的
undervalue [͵ʌndə'vælju:] *v.* 低估
undulating ['ʌndjə͵letiŋ] *adj.* 波状的；起伏的
undulation [͵ʌndju'leiʃən] *n.* 波动
uniformity [͵ju:ni'fɔ:miti] *n.* 一致；一式；均匀
unlikely [ʌn'laikli] *adj.* 未必的；不太可能的；靠不住的
unprojected [͵ʌnprəu'dʒektid] *adj.* 未计划的；非预料的
unreserved ['ʌnri'zə:vd] *adj.* 不隐瞒的；坦白的；无限制的
unswerving [ʌn'swə:viŋ] *adj.* 不歪的；不偏离的；坚定的；始终不渝的
ultraviolet [͵ʌltrə'vaiəlit] *adj.* 紫外的；紫外线的；利用紫外线的 *n.* 紫外光；紫外辐射
upload ['ʌpləud] *v. & n.* 上载
urban ['ə:bən] *adj.* 城市的；市内的
usability [͵ju:zə'biləti] *n.* 可用性
validity [və'liditi] *n.* 有效性；合法性；正确性

valuation [ˌvæljuˈeiʃən] *n.* 估价；评价；计算
vector [ˈvektə] *n.* [数]向量；矢量；带菌者
vegetation [ˌvedʒiˈteiʃən] *n.* [植]植被；(总称)植物；草木；(植物的)生长；呆板单调的生活
vehicle [ˈviːikl] *n.* 交通工具；车辆；媒介物；传达手段
versus [ˈvəːsəs] *prep.* 与……相对，对抗(指诉讼，比赛中)
vibration [vaiˈbreiʃən] *n.* 振动；颤动；摇动；摆动
vicinity [viˈsiniti] *n.* 邻近；附近；接近
viewshed [ˈvjuːʃed] *n.* 可视域
visible [ˈvizibəl] *adj.* [物理][光]可见的；看得见的；明显的 *n.* 看得见的事物
visibility [ˌviziˈbiliti] *n.* 可见度；能见度；可见性；显著；明显度
vista [ˈvistə] *n.* 展望；回想；街景
visualization [ˌvizjuəlaiˈzeiʃən] *n.* 可视化；清楚地呈现
volcanism [ˈvɔlkənizəm] *n.* 火山作用
volcanology [ˌvɔlkəˈnɔlədʒi] *n.* 火山学
voltage [ˈvəultidʒ] *n.* [电工]电位；电压；伏特数
watershed [ˈwɔːtəʃed] *n.* 分水岭；流域
wearable [ˈwerəbl] *adj.* 可穿用的；可佩带的
    wearable technology 可穿戴技术
workshop [ˈwəːkʃɔp] *n.* 研讨会
wrap [ræp] *vi.* 缠绕；重叠；穿外衣；包起来
wriggle [ˈrigl] *n. & v.* 扭动；蠕动
yaw [jɔː] *n.* (火箭、飞机、宇宙飞船等)偏航
zenith [ˈzeniθ] *n.* 天顶；顶点；顶峰；最高点
zoom [zuːm] *vi.* 突然扩大；急速上升；摄象机移动

Appendix II Glossary of Terms

absolute error 绝对误差

absolute gravity measurement 绝对重力测量

absolute orientation 绝对定向

absolute positioning 绝对定位

acceleration of gravity 重力加速度

active microwave sensors 主动微波遥感传感器

active remote sensing 主动式遥感

addition constant 加常数

adjusted value 平差值

adjustment of correlated observation 相关平差

adjustment of observations, survey adjustment 测量平差

adjustment of typical figures 典型图形平差

aerial photogrammetry 航空摄影测量

aerial photography 航空摄影

aerial triangulation 空中三角测量

airborne remote sensing 航空遥感；机载遥感

airborne surveying 航空测量

alignments survey 定线测量

ambiguity resolution 模糊度解算

analytical aerotriangulation 解析空中三角测量；电算加密

analytical photogrammetry 解析摄影测量

analytical plotter 解析测图仪

angle closing error of traverse 导线角度闭合差

annexed leveling line 附合水准路线

approximate adjustment 近似平差

arbitrary projection 任意投影

artificial Earth satellite 人造地球卫星

as-built survey 竣工测量

ascending node 升交点

astronomic positioning 天文定位

atomic clock 原子钟

attribute data 属性数据

automatic level, compensator level 自动安平水准仪

automatic target recognition（ATR）目标自动识别
average error 平均误差
azimuthal projection 方位投影
backsight（BS）后尺
barometric leveling 气压水准测量
base station 基站
basic traverse 基本导线
BeiDou Navigation Satellite System（BDS）北斗导航卫星系统［中］
block adjustment 区域网平差
BM(benchmark) 水准基点
breakthrough survey 贯通测量
breakthrough tolerance 贯通限差
breakthrough error 贯通误差
bridge survey 桥梁测量
broadcast ephemeris 广播星历
buffering analysis 缓冲区分析
building axis survey 建筑轴线测量
building engineering survey 建筑工程测量
bundle adjustment 光束法平差
carrier phase correction 载波相位改正
carrier phase smoothed pseudorange method 载波相位平滑伪距方法
CCD camera CCD 相机
celestial body 天体
celestial coordinate system 天球坐标系
centrifugal force 离心力
chain code 链码
CIO(conventional international origin) 国际协议原点
circular encoders 编码度盘
city survey 城市测量
clock error 钟差
closed leveling line 闭合水准路线
closed loop traverse 闭合环导线
closed traverse 闭合导线
close-range photogrammetry 近景摄影测量
closing error in coordinate increment 坐标增量闭合差
Coarse/Acquisition Code（C/A code）C/A 码
collimation line method 视准线法
combined adjustment 联合平差
command tracking station(CTS) 指令跟踪站
computer graphics 计算机图形学

conditional adjustment with parameters 附参数条件平差
conditional equation 条件方程
conditional adjustment 条件平差
conformal projection 等角投影；正形投影
connecting traverse 附合导线
constant error 常差
construction control network 施工控制网
construction plan 施工平面图
construction survey 施工测量
construction traverse 施工导线
Continuously Operating Reference Station（CORS）连续运行参考站系统
control network 控制网
control network for deformation observation 变形观测控制网
control point 控制点
control segment 控制部分
control survey 控制测量
coordinate conversion 坐标变换
Coordinated Universal Time（UTC）世界协调时；世界统一时间
correlation survey 联系测量
covariance function 协方差函数
cross section 横断面，断面图，剖面图
cross-section survey 横断面测量
crust deformation measurement 地壳形变观测
crustal deformation 地壳变形
crustal motion 地壳运动
CSGPC（Chinese Society of Geodesy, Photogrammetry and Cartography）中国测绘学会
cut-and-cover method 明挖法
data analysis 数据分析
data acquisition（capture）数据获取(采集)
data classification 数据分类
data compression 数据压缩
data display 数据显示
data exploration 数据探查
data management 数据管理
data recorder 电子手簿，数据采集器
data transfer 数据转换
database management system（DBMS）数据库管理系统
deflection observation 挠度观测
deformation monitoring(observation) 变形监测(观测)
dense image matching 影像密集匹配

depression angle 俯角；俯视角

detail survey 碎部测量

differential correction 差分改正

differential GNSS (DGNSS) 差分 GNSS

differential leveling 微差水准测量

digital elevation model (DEM) 数字高程模型

digital line graph (DLG) 数字线划图

digital orthoimage 数字正射影像

digital orthophoto map (DOM) 数字正射影像图

digital photogrammetry 数字摄影测量

digital surface model (DSM) 数字表面模型

digital terrain model (DTM) 数字地面模型

digital terrain visualization 数字地形可视化

direct adjustment 直接平差

direct leveling, spirit leveling 几何水准测量

direct plummet observation 正垂线观测

displacement observation 位移观测

distance measurement 距离测量

distance measuring instrument, rangefinder 测距仪

distance-measuring error 测距误差

distortion coefficients 畸变系数

drill and blast excavation method 钻爆开挖法

dual-frequency 双频

dynamic height 力高

Earth Centered Earth Fixed (ECEF) coordinate system 地心地固坐标系

Earth resources technology satellite-1 (ERTS-1) 地球资源技术卫星 1 号

earth tide 地球潮汐；陆潮

Earth's flattening 地球扁率

eccentricity of ellipsoid 椭球偏心率

EDM(electronic distance measurement) 电子测距仪

electromagnetic distance measuring instrument 电磁波测距仪

electromagnetic radiation 电磁辐射

electromagnetic spectrum 电磁波频谱；电磁波谱；电磁光谱

electronic level 电子水准仪

electronic theodolite 电子经纬仪

electro-optical distance measuring instrument 光电测距仪

elevation angle 高度角

elevation difference 高差

elevation of sight 视线高程

engineering control network 工程控制网

engineering survey 工程测量
equidistant projection 等距投影
equivalent projection 等积投影
error distribution 误差分布
error ellipse 误差椭圆
error of closure, closing error, closure 闭合差
error of focusing 调焦误差
error propagation, propagation of error 误差传播
error test 误差检验
expectation, expected value 期望值
exterior orientation parameters 外方位参数
fault detection 故障检测
FIG(Federation of International Surveyors) 国际测量师联合会
figure of the Earth 地球形状
fissure observation 裂缝观测
fixed error 固定误差
flattening [of ellipsoid] 椭球扁率
flying platform 飞行平台
flight altitude 航高
flight control system 飞行控制系统
foresight (FS) 前尺
forward intersection 前方交会
free station 自由设站法
full digital (softcopy) photogrammetry 全数字摄影测量；软拷贝摄影测量
functional model 函数模型
GALILEO 伽利略系统(欧盟)
Gaussian distribution 高斯分布；正态分布
geodesy 大地测量学
geodetic astronomy 大地天文学
geodetic azimuth 大地方位角
geodetic coordinate 大地坐标
geodetic coordinate system 大地坐标系统
geodetic datum 大地基准
geodetic height, ellipsoidal height 大地高
geodetic instrument 大地测量仪器
geodetic latitude 大地纬度
geodetic longitude 大地经度
geodetic network 大地网
geodetic origin 大地原点
geodetic position 大地位置

Geodetic Reference System 1980（IRS8O）1980 国际大地参考系统
geodetic reference system of 1980（GRS 80）1980 大地测量参考系
geodetic surveying 大地测量；大地测量学
geodimeter 光速测距仪；光电测距仪
geographic information systems（GIS）地理信息系统
geoid 大地水准面
geoidal height, geoid undulation(N) 大地水准面高；大地水准面差距
geoidal undulation 大地水准面高
geological survey 地质测量
geomatics 测绘学
geometric geodesy 几何大地测量学
georelational data model 地理关系数据模型
Geo-robot 测量机器人
geospatial data 地理空间数据
geo-synchronous satellite 地球同步卫星
GIS(geographic information system) 地理信息系统
GIS modeling GIS 建模
Global Navigational Satellite System（GNSS）全球导航卫星系统
GLObal NAvigation Satellite System（GLONASS）GLONASS 导航卫星系统(俄)
Global Positioning System（GPS）GPS 定位系统(美)
gnomonic projection 球心投影；极平投影
GNSS constellation GNSS 星座
GNSS receiver GNSS 接收机
gravimeter, gravity meter 重力仪
gravimetric baseline 重力基线
gravimetric deflection 重力偏差
gravimetric deflection of the vertical 重力垂线偏差
gravimetric leveling 重力水准测量
gravitational constant 引力常数；重力常数
gravitational field 重力场；引力场
gravitational potential 引力位
gravity anomaly 重力异常
gravity datum 重力基准
gravity field 重力场
gravity gradient measurement 重力梯度测量
gravity measurement 重力测量
gravity observation of Earth tide 重力固体潮观测
gravity potential 重力势，重力位
gravity reduction 重力归算
gravity station 重力点

gravity vector 重力矢量
grid bearing 坐标方位角
gross error 粗差
gross error detection 粗差检验
ground-based control complex(GCS) 地面控制部分
ground control system, ground control station (GCS) 地面控制系统，地面控制站
gyro azimuth 陀螺方位角
gyroscopic theodolite 陀螺经纬仪
height anomaly 高程异常
height of instrument (HI) 仪器高
height of target (HT) 目标高
histogram 直方图
homologous points, corresponding image points 同名像点
homologous rays 同名光线，同名射线
horizontal angle 水平角
horizontal circle 水平刻度盘
horizontal control network 平面控制网；水平控制网
horizontal refraction error 水平折光差
horizontal survey 水平测量；平面测量
Huanghai vertical datum of 1956 1956黄海高程系统
hydrographic engineering survey 水利工程测量
hydrographic survey 水道测量
hyperspectral image (HSI) 高光谱图像
IAG (International Association of Geodesy) 国际大地测量协会
IAM (International Society of Mine Surveying) 国际矿山测量协会
ICA (International Cartographic Association) 国际制图协会
IGU (International Geographical Union) 国际地理联合会
IHO (International Hydrography Organization) 国际海道测量组织
image cube 图像立方体
image data 图像数据
image mosaicing 影像镶嵌
image overlap 影像重叠
image point 像点
image rectification 影像纠正
immersed tube (sunken tube) tunnel 沉管法
independent coordinate system 独立坐标系
index error of vertical circle 竖盘指标差
index of precision 精度指标
indirection adjustment 间接平差
inertial navigation system (INS) 惯性导航系统

inertial measurement unit (IMU) 惯性测量单元
infrared EDM instrument 红外测距仪
infrared remote sensing 红外遥感
instrumental error 仪器误差
intensity value 强度值，亮度值
internal orientation 内定向
interior parameters 内参
International GNSS Service (IGS) 国际 GNSS 服务
interferometric synthetic aperture radar (InSAR) 干涉合成孔径雷达
interometry SAR 干涉雷达
inverse of weight matrix 权逆阵
inverse plummet observation 倒垂线观测
ionospheric delay 电离层延迟
ISDE (International Society for Digital Earth) 国际数字地球协会
ISPRS (International Society for Photogrammetry and Remote Sensing) 国际摄影测量与遥感学会
IUGG (International Union of Geodesy and Geophysics) 国际大地测量与地球物理联合会
kinematic positioning 动态定位
kinematic relative positioning 动态相对定位
kinematic single point positioning 动态单点定位
Krasovsky ellipsoid 克拉索夫斯基椭球
Lambert projection 兰勃特投影
land management 土地管理
land survey, property survey, boundary survey, cadastral survey 土地测量，地籍测量
laser distance measuring instrument, laser ranger 激光测距仪
laser level 激光水准仪
law of probability 概率论
law of universal gravitation 万有引力定律
laws of probability 概率论
least-cost path analysis 最小耗费路径分析
least square method 最小二乘法
least squares collocation 最小二乘配置法；最小二乘拟合推估法
least-squares adjustment 最小二乘平差
level 水准仪
level rod 水准尺
light detection and ranging (LiDAR) 激光雷达
limit error 极限误差
line feature 线特征
linear intersection 边交会法
linear-angular intersection 边角交会法

local navigation satellite system 区域导航卫星系统

local operation 局域运算

long-range EDM instrument 远程电子测距仪

lunar laser ranging(LLR) 激光测月

magnetic azimuth 磁方位角

map projection 地图投影

marine survey 海洋测量

master control station 主控站

mean sea level（MSL）平均海[水]面

mean square error of a point 点位中误差

mean square error of angle observation 测角中误差

mean square error of azimuth 方位角中误差

mean square error of coordinate 坐标中误差

mean square error of height 高程中误差

mean square error of side length 边长中误差

mean square error(MSE) 中误差

medium Earth orbit（MEO）中地球轨道

Mercator projection 墨卡托投影

method by series，method of direction observation 方向观测法

method in all combinations 全组合测角法

method of laser alignment 激光准直法

method of tension wire alignment 引张线法

microwave distance measuring instrument 微波测距仪

microwave remote sensing 微波遥感

mine survey 矿山测量

mining subsidence observation 开采沉陷观测

minor angle method 小角度法

mobile mapping system（MMS）移动测量系统

mobile mapping technology（MMT）移动测量技术

monitor station 监控站

most probable value(MPV) 最或然值

multipath effect 多路径效应

multiplication constant 乘常数

national vertical datum of 1985 1985国家高程基准

navigation message 导航电文

navy navigation satellite system(NNSS) 海军导航卫星系统

near infrared（NIR）近红外

neighborhood operation 邻域运算

network analysis 网络分析

network of multiple reference stations 多基站网络

network RTK 网络 RTK 技术
nomal projection 正轴投影
nominal accuracy 标称精度
non-photographic remote sensing 非摄影方式遥感
normal distribution 正态分布
normal equation 法方程
normal error distribution curve 正态误差分布曲线
normal height 正常高
normal random variable 正态随机变量
object-based data model 基于对象的数据模型
oblique observation, tilt observation 倾斜观测
oblique photogrammetry 倾斜摄影测量
oblique projection 斜轴投影
observation equation 观测方程
observation error 观测误差
observation of slope stability 边坡稳定性观测
open traverse 支导线
optical level 光学水准仪
optical plummet 光学对中器
optical theodolite 光学经纬仪
orthogonal projection 正交投影；正射投影
orthographic projection 正射投影
orthometric heights(H) 正高
orthophoto 正射像片
ortho-rectification 正射纠正
overlay analysis 叠置分析
pacing 步测；定步
pass point 加密点
parametric adjustment with conditions 附条件参数平差；附条件间接平差
parametric adjustment 参数平差
passive positioning system 被动式定位系统
passive remote sensing 被动式遥感，无源遥感
personal error 人为误差
perspective projection 透视投影
photo tilt 像片倾斜
photogrammetry 摄影测量学
photographic principal distance 摄影主距
photographic remote sensing 摄影方式遥感
photographic scale 摄影比例尺
physical distance measure operation 自然距离量测运算

physical geodesy 物理大地测量学；大地重力学
picture element/ pixel 像元/像素
pipe survey 管道测量
pitch, roll, and yaw/heading 俯仰；横滚；航偏
plane surveying 平面测量；平面测量学
planetary geodesy 行星大地测量学
point feature 点特征
polar motion 极移
polyconic projection 多圆锥投影
polygon（area）feature 面特征
Positioning, Navigation and Timing（PNT）定位导航和授时
position and orientation system（POS）定位定向系统
POS-assisted aerial triangulation POS辅助空中三角测量
positive positioning system 主动式定位系统
post-processed differential correction 后处理差分改正
precise alignment 精密准直
precise code 精码
precise ephemeris 精密星历
Precise Point Positioning（PPP）精密单点定位
Precise Positioning Service（PPS）精密定位服务
precise ranging 精密测距
precision 精度
preliminary survey 初测
primary traverse 主要导线
probability density function 概率密度函数
probable error 或然误差
profile diagram, profile 纵断面图
profile survey 纵断面测量
property line survey 建筑红线测量
proportional error 比例误差
pseudorange 伪距
pssive remote sensing 被动式遥感
pssive/active microwave sensing 被动微波遥感
public engineering survey 市政工程测量
quasi-geoid 似大地水准面
quasi-stable adjustment 拟稳平差
radio detection and ranging（Radar）雷达
random error, accident error 随机（偶然）误差
rank defect adjustment 秩亏平差
raster data 栅格数据

raster data model 栅格数据模型
real-time differential correction 实时差分改正
Real-Time Kinematic (RTK) positioning 实时动态定位
real-time positioning 实时定位
receiver antenna 接收机天线
rectangular grid 直角坐标网
redundant observation 多余观测
reference datum 参考基面；参考基准面
reference ellipsoid 参考椭球
reference receiver 基准接收机
reference station 基站
reflecting stereoscope 反光立体镜
refraction correction 折光差改正
regular grid 规则格网
relational database management system (RDBMS) 关系数据库管理系统
relative error 相对误差
relative gravity measurement 相对重力测量
relative orientation 相对定向
relative positioning 相对定位
reliability 可靠性
remote controller 远距离控制器
remote sensing(RS) 遥感
remote sensor 遥测传感器；遥感器
resection 后方交会
right ascension 赤经
rigorous adjustment 严密平差
river-crossing leveling 跨河水准测量
road engineering survey 道路工程测量
robotic (motorized) total station 智能型全站仪，测量机器人
rotation parameters 旋转参数
route survey 路线测量
roving receiver 流动接收机
roving station 流动站
run-length encoding 游程编码
satellite clock 卫星钟
satellite geodesy 卫星大地测量学
satellite laser ranger 卫星激光测距仪
satellite laser ranging(SLR) 卫星激光测距
satellite positioning 卫星定位
satellite ranging 卫星测距

satellite remote sensing 卫星遥感
satellite-to-satellite tracking 卫星跟踪卫星技术
scale invariant feature transform（SIFT）尺度不变特征变换
scale parameter 尺度参数
sea surface topography（SST）海面地形
Selective Availability（SA）选择可用性
selenodesy 月面测量；月面测量学
semi-major axis [of ellipsoid] 椭球长半轴；地球长半轴
sensitivity 灵敏度
sensor system 传感器系统
sensor parameters 传感器参数
sequential adjustment 序贯平差
servo motors 伺服马达
setting-out of main axis 主轴线放样
setting-out survey, construction layout 施工放样
settlement（subsidence）observation 沉陷观测
shield tunneling method 盾构法
side intersection 侧方交会
side-looking airborne radar system（SLAR）机载侧视雷达系统
sighting distance 视距
single point positioning 单点定位
single reference station 单基站
site map 工地(总)平面图
space geodesy 空间大地测量学
space segment 空间部分
sparse reconstruction 稀疏重建
spatial analysis 空间分析
spatial interpolation 空间插值
spur leveling line 支水准路线
stadia addition constant 视距加常数
stadia hair 视距丝；视距线
stadia interval 视距间隔
stadia multiplication constant 视距乘常数
standard deviation 标准差
standard field of length 长度标准检定场
Standard Positioning Service（SPS）标准定位服务
static positioning 静态定位
static point positioning 静态单点定位
static relative positioning 静态相对定位
stereo glasses 立体眼镜

stereocomparator 立体坐标量测仪
stereopair, stereo photopair 立体像对
stereophotogrammetry 立体摄影测量
stochastic model 随机模型
surface control survey 地面控制测量
surface-fitting algorithm 曲面拟合算法
survey of present state at industrial site 工厂现状图测量
survey specifications, specifications of surveys 测量规范
surveying and mapping 测绘
surveying for site selection 厂址测量
synthetic aperture radar (SAR) 合成孔径雷达
system control center(SCC) 系统控制中心
systematic error 系统误差
tacheometry, stadia 视距测量(法)
terrain analysis 地形分析
terrestrial remote sensing 地面遥感
theory of error 误差理论
thermal infrared (TIR) 热红外
thermal infrared detector 热红外探测器
tie points matching 连接点匹配
tolerance 限差
topographic survey 地形测量
topological relationship 拓扑关系
total length closing error of traverse 导线全长闭合差
total station 全站仪
translation parameters 平移参数
transverse projection 横轴投影
traverse angle 导线折角
traverse leg 导线边
traverse network 导线网
traverse point 导线点
traversing 导线测量
triangular irregular network (TIN) 不规则三角网
triangulateration 边角测量
triangulateration network 边角网
triangulation 三角测量
triangulation network 三角网
trigonometric leveling 三角高程测量
trilateration 三边测量
trilateration network 三边网

tropospheric delay 对流层延迟
true error 真误差
true north 真北
tunnel boring machine (TBM) 隧道掘进机
tunnel construction methods 隧道施工方法
tunnel guidance system 隧道引导系统
tunnel lining 隧道衬砌
tunnel survey 隧道测量
two-color laser ranger 双色激光测距仪
ultraviolet remote sensing 紫外遥感
unbiased estimate 无偏估计
underground control survey 地下控制测量
Universal Polar Stereographic projection(UPS) 通用极球面投影
Universal Transverse Mercator (UTM) 通用横墨卡托投影
unmanned aerial vehicle (UAV) 无人机；无人飞行器
unmanned aircraft system (UAS) 无人机系统
up-link station 注入站
user segment 用户部分
variance 方差
variance of unit weight 单位权方差；方差因子
variance-covariance matrix 方差-协方差矩阵
variance-covariance propagation law 方差-协方差传播律
varioscale projection 变比例投影
vector data 矢量数据
vector data model 矢量数据模型
vertical angle 垂直角
vertical circle 垂直度盘
vertical control network 高程控制网
vertical photograph 垂直像片
vertical photogrammetry 垂直摄影测量
vertical survey 高程测量；垂直测量
vertical shaft 竖井
very long baseline interferometry(VLBI) 甚长基线干涉测量
viewshed analysis 视域分析
visible light remote sensing 可见光遥感
watershed analysis 流域分析
weight coefficient 权系数
weight function 权函数
weight reciprocal of figure 图形权函数
weighting matrix 权矩阵

world geodetic system 1984（WGS-84） WGS-84 世界大地坐标系
wriggle survey 收边测量；断面测量
Xi'an Geodetic Coordinate System 1980 1980 西安坐标系
zenith angle 天顶距
zenith distance 天顶距
zonal operation 分区运算

References

1. 测绘学名词审定委员会. 测绘学名词[M]. 4版. 北京：测绘出版社，2020.
2. 胡庚申主编. *English Paper Writing & Publication*[M]. 北京：高等教育出版社，2000.
3. 杨国东，王民水. 倾斜摄影测量技术应用及展望[J]. 测绘与空间地理信息，2016，39(01)：13-15，18.
4. 闫利. 低空无人机遥感技术与应用[M]. 武汉：武汉大学出版社，2022.
5. 阎庆甲，阎文培编著. 科技英语翻译方法[M]. 北京：冶金工业出版社，1992.
6. 王泉水. 科技英语翻译技巧[M]. 天津：天津科学技术出版社，1989.
7. 王佩军，徐亚明. 摄影测量学[M]. 3版. 武汉：武汉大学出版社，2016.
8. 张祖勋，张剑清. 数字摄影测量学[M]. 2版. 武汉：武汉大学出版社，2012.
9. Aber J. S., Marzolff I., Ries J. *Small-format Aerial Photography: Principles, Techniques and Geoscience Applications*[M]. Elsevier, 2010.
10. ACSM (2004) Journal of Surveying Engineering. http://scitation.aip.org/suo/
11. Aerospace (2023) A brief history of GPS, [2023-03-18]. https://aerospace.org/article/brief-history-gps
12. Albertz J. A Look Back[J]. *Photogrammetric Engineering & Remote Sensing*, 2007, 73(5): 504-506.
13. AGU (2004) Publication. http://www.agu.org/pubs/pubs.html#journals.
14. ASPRS. WHAT IS ASPRS? [EB/OL]. (2015-05-20)[2022-10-10]. https://www.asprs.org/organization/what-is-asprs.html
15. Barry F. Kavanagh. *Geomatics*[M]. Pearson Education Inc., 2003.
16. Barry F. Kavanagh and S. J. Glenn Bird. *Surveying: Principles and Applications*[M]. (4th Ed.) Prentice-Hall Inc., 1996.
17. Boon M. A. Comparison of a fixed-wing and multi-rotor UAV for environmental mapping applications: A case study[J]. *Environment Sicence, Engineering*, 2017.
18. CASLE (2004) Survey Review. http://www.surveyreview.org/index.html
19. Caulfield, J. The Four Main Types of Essay | Quick Guide with Examples[EB/OL]. (2021-12-06)[2023-03-13]. Scribbr. https://www.scribbr.com/academic-essay/essay-types/
20. Chang, Kang-Tsung. *Introduction to Geographic Information Systems* (9th Ed.)[M]. McGraw-Hill Education, 2018.
21. CIG (2004) GEOMATICA. (http://www.cig-acsg.ca)
22. Course. CV vs. Resume: What's the Difference? | Coursera May 5, 2022.
23. C. P. Lo & Albert K. W. Yeung. *Concepts and Techniques of Geographic*

Information Systems[M]. Prentice-Hall, Inc., 2002.

24. DIT (2002) Geomatics Background Information. http://www.dit.ie/DIT/built/geomatics/whatisgeomatics.html
25. Elsevier (2004) Journal of Geodynamics. http://www.elsevier.com/wps/find/journaldescription.cws_home/874/description
26. ESRI (2022) What is GIS? [EB/OL]. [2022-12-10]. https://www.esri.com/en-us/what-is-gis/overview
27. Fahlstrom P. G., Gleason T. J., Sadraey M. H. *Introduction to UAV Systems*[M]. John Wiley & Sons, 2022.
28. FIG (2004) About FIG. http://www.fig.net/general/leaflet-english.htm
29. Flatsurv (2002) Basic theories of surveying. http://www.flatsurv.com/
30. Francts H. Moffitt and John D. Bossler. *Surveying*(10th Ed.)[M]. Addison Wesley Longman, Inc., 1998.
31. Geography Matters™ An ESRI © White Paper (2002). http://www.gis.com/whatisgis/geographymatters.pdf
32. González-Jorge H., Martínez-Sánchez J., Bueno M, et al. Unmanned aerial systems for civil applications: A review[J]. *Drones*, 2017, 1(41): 2.
33. GPS World (2004) The Magazine. http://www.gpsworld.com/gpsworld/
34. Grafarend E., Sanso F. *Optimization and Design of Geodetic Networks*[M]. Elsevier B. V., 1985.
35. GSD (2003) Geodesy. http://www.geod.nrcan.gc.ca/
36. Heipke C., Madden M., Li Z, et al. Theme issue: State-of-the-art in photogrammetry, remote sensing and spatial information science[J]. *ISPRS Journal of Photogrammetry and Remote Sensing*, 2016, 115: 1-2.
37. ICA (1995) Mission. http://www.icaci.org/
38. IHO (2004) Background. http://www.iho.shom.fr/
39. IUGG (2004) About IUGG. http://www.iugg.org/eoverview.html
40. IAG (2003) IAG Organization. http://www.iag-aig.org
41. ISPRS (2004) Objectives and Activities. http://www.isprs.org/society.html
42. James M. Anderson and Edward M. Mikhail. *Surveying: Theory and Practice*(7th Ed.)[M]. WCB/McGraw Hill Company, 1998.
43. Jennifer Herrity. How to Write a Cover Letter (With Examples and Tips). [EB/OL]. (2023-03-11) [2023-03-16]. http//www.indeed.com/career-advice/resume-cover-letters/how to write a cover letter/
44. Jennifer Herrity. Resume vs. CV (Curriculum Vitae): Key Document Differences. [EB/OL]. (2023-03-11) [2023-03-16]. http//www.indeed.com/career-advice/resume-cover-letters/difference-between-resume-and-cv/
45. Liang S, Wang J. *Advanced Remote Sensing: Terrestrial Information Extraction and Applications*[M]. Academic Press, 2019.
46. National Academy of Sciences (1978) Commission on Geodesy — Geodesy: Trend

and prospects. National Academy of Sciences, Washington, D. C.

47. National Geographic Education. GIS (Geographic Information System) [EB/OL]. (2022-12-20) [2023-1-10]. https://education.nationalgeographic.org/resource/geographic-information-system-gis/

48. Office of the DoD (2020) Global Positioning System Standard Positioning Service Performance Standard (5th Edition). https://www.gps.gov/technical/ps/2020-SPS-performance-standard.pdf.

49. On-Line Resources for Land Surveying and Geomatics (2001). http://surveying.mentabolism.org/

50. Paul R. Wolf and Charles D. Ghilani. *Adjustment Computations: Statistics and Least Squares in Surveying and GIS*[M]. John Wiley & Sons, Inc., 1997.

51. Petrie G., and Kennie T. J. M. Terrain Modelling in Surveying and Civil Engineering, Caithness: Whittles Publishing in Assoc. with Thomas Talford, 1990: 1-2.

52. Petr Vanicek and Edward J. Krakiwsky. *Geodesy: the Concepts* (2nd Ed.) [M]. Elsevier Science Publish Company, Inc., 1986.

53. PIX4D. Diving into underwater photogrammetry [EB/OL]. (2017-04-20) [2022-10-10]. https://www.pix4d.com/blog/diving-into-underwater-photogrammetry/.

54. Rizos, Chris; Volker Janssen, Craig Roberts and Thomas Grinter. Precise Point Positioning: Is the Era of the Differential GNSS Positioning Drawing to an End? FIG Working Week 2012. Knowing to manage the territory, protect the environment, evaluate the cultural heritage. Rome, Italy, 6-10, 2012.

55. Schowengerdt R. A. *Remote Sensing: Models and Methods for Image Processing*[M]. Elsevier, 2006.

56. Sciencedirect (2004) Journals. http://www.sciencedirect.com/science/journal/09242716

57. SCO (2004) GIS Basics. http://www.geography.wisc.edu/sco/gis/basics.html

58. SNP (2000) Notes on Basic GPS Positioning and Geodetic Concepts. http://www.gmat.unsw.edu.au/snap/gps/gps_notes.htm

59. Springerlink (2004) Publication. http://link.springer.de/link/service/journals/00190/index.htm

60. Taylor & Francis (2004) Journals. http://www.tandf.co.uk/journals/titles/13658816.asp

61. Toth C., Jóźków G. Remote sensing platforms and sensors: A survey[J]. *ISPRS Journal of Photogrammetry and Remote Sensing*, 2016, 115: 22-36.

62. University of Calgary. Faculty of Graduate Studies[EB/OL]. [2022-12-30]. https://grad.ucalgary.ca/future-students.

63. University of Waterloo. Centre for Career Action[EB/OL]. (2011-01-24) [2023-01-10]. https://uwaterloo.ca/career-action/

64. USGS. What is remote sensing and what is it used for? [EB/OL]. [2005-04-06; 2022-10-10]. https://www.usgs.gov/faqs/what-remote-sensing-and-what-it-used

65. University of Melbourne (2004). Introduction to Geomatics. http: //www. sli. unimelb. edu. au/planesurvey/prot/topic/topic. html
66. U. S. Department of Defense. *Unmanned Aircraft Systems Roadmap* 2005—2030 [M]. Lulu Press, 2015.
67. Wanninger, Lambert (2008): Introduction to Network RTK, IAG Working Group 4.5.1: Network RTK, http: //www. wasoft. de/e/iagwg451/intro/introduction. html
68. Wang Zhizhuo. *Principles of Photogrammetry (with Remote Sensing)* [M]. Publishing House of Surveying and Mapping, 1990.
69. Wolf P. R., Dewitt B. A., Wilkinson B. E. *Elements of Photogrammetry with Applications in GIS*[M]. McGraw-Hill Education, 2014.
70. W. Schofield. *Engineering Surveying: Theory and Examination Problem for Students*(5th Ed.)[M]. Butterworth-Heinemann, 2001.
71. Yin Hui and Wang Jianguo. *Engineering Surveying (in English)*[M]. Wuhan University Press, 2022.